CB Radio

John Gordon
G7WG1

RADIO ACTIVE

02392 313093

Let me introduce myself . . .

I have been in the communications business for fifteen years and selling CB equipment almost every day of that period.

Customers often ask me "What books have you got on CB Radio"? I have to answer "None" because there really aren't any anymore; those that are around are five or six years old. Because of this I could see a need for information in some form, so I thought I would put together 4 or 5 pages of literature as a guide for the new "breaker". Well! those 4 or 5 pages got bigger and bigger until one day someone said to me "Why don't you make a book of it", and so I added more and more and this is the end product.

I have tried to cover all the aspects of Citizens Band Radio and I hope that in some way it is of some help to you. I was the first person to use CB in my city, in the illegal days, but I have deliberately avoided including AM and SSB because in my line of work I cannot be seen to condone the use of it, I hope you "Aunty Mary" users will understand.

John Gordon

(The Spirit)

Radio Active Publications

Published by

1A Munster Road
North End
Portsmouth
Hampshire
PO2 9BS

ACKNOWLEDGEMENTS

PAUL MARTIN **(G0AFF)**

KEVIN O' BRIEN **(G1MZU)**

MIKE DEVEREUX **(G3SED)**

MICK HONEYWELL **(G0ABB)**

QSK the STEAM QUEEN

CHANNEL 07 BREAKERS

CTE INTERNATIONAL

NEVADA COMMUNICATIONS

RADIO COMMUNICATIONS AGENCY
(JULIA FRASER)

SIRIO ANTENNAS

TEAM CB RADIO

ZETAGI (ITALY)

ALEX NEWBERRY
Graphic design

All my CUSTOMERS for their input

My wife MARGARET
for the supply of coffee and letting me commandeer the computer.

Thanks to everyone else who helped make this book possible.

CONTENTS. . . . .

Chapter 1

CB UP TO-DATE

It was in the late 1970's that CB radio first found its way to the shores of the UK. Travellers to the USA had seen low cost CB radio equipment in widespread use across the States. Many returned home with the radios and began using them on illegal frequencies in the UK. At that time they faced confiscation and heavy fines if they were caught in possession of the radios - many were unaware of the serious nature of using radio illegally. It wasn't long before the trickle became an avalanche and CB users formed clubs that started to campaign and fight for the legalisation of CB radio in the UK. Surprisingly, it was just a few years, through the mounting pressure of these groups, before CB was finally made legal in November 1981. The system adopted allowed 40 channels (with FM modulation on 27.60125 - 27.99125 MHz). The previously imported American equipment used AM and SSB modulation on frequencies that were not allowed in the UK. To further complicate matters, the 40 channels allocated to the UK did not conform with any other European country. In an effort to give Europe a common system, a second band of 40 channels was allocated in December 1986 -these were on the frequency range 26.965 - 27.405 MHz (FM Mode).

Why is CB so popular? Probably the main reason is that it supplies the user with instant communication, anywhere. Furthermore it is useful, inexpensive, easy to use and fun. CB is used daily by truckers, housewives, farmers, motorists, in fact, people from all walks of life. It can turn a humdrum motoring trip into a pleasant journey. CB radio gets you where you want to go, brings help when needed, provides company on a long journey and helps you obtain traffic reports. CB allows a person to talk and even help another without personal contact. CB is the perfect way to get involved in and yet still remain apart and anonymous from a particular situation at the touch of a microphone. CB can be vital in an emergency; reports of rescues involving the use of CB have been numerous.

Today there is probably not one country in the world where CB is not used in one form or another. In countries like Africa as a communication between villages the cheapness and simplicity of 27 MHz means that it is ideal. There are lots of terrific "breakers" to meet out there, so if you haven't already got a CB rig, get one and get "on the air" today.

FACT-FILE

CB first started in the U.S.A. in 1945 but was not a success. Two years later in April 1947 they were allocated the frequency of 467 MHz. This was launched with 50 channels on FM with 50 kHz spacing.

It wasn't until 1958 that The Citizens Band Radio Service was established in the U.S.A. to provide a short-distance radio communication service. This radio service, better known as "Citizens Band" was a 23 channel radio around the 27 megahertz band (26.960-27.280 MHz) in AM mode. In the early days, a CB set was heavy, bulky and often drained the battery of the vehicle in which it was installed. The advent of the transistor reduced the size and the weight of the set, and the trucker's strike, fuel shortages and 55 m.p.h. speed limit of the early 70's contributed to the skyrocketing popularity of CB radio. Truckers could easily find where fuel was available and could warn one another about speed traps. Because of its popularity, in 1976 the U.S.A. increased its CB channels to 40, but still on AM. In that same year 10 million CB sets were sold in the U.S.A. The re-allocation of 40 channels was accompanied by a rule that all 23 channel CB Radios must be off the shelves of any shop by 1st of January 1977. There was a panic sale with all the old 23 channel sets virtually being given away. Today it is estimated that there are over 50 million Americans using CBRadios.

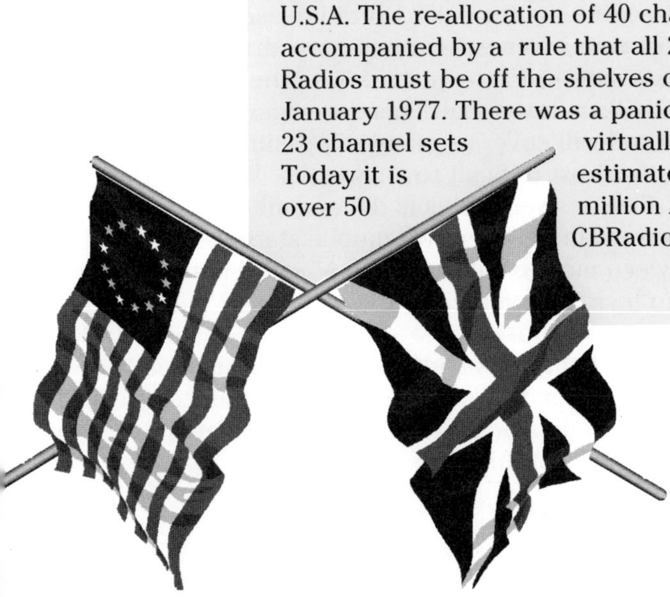

Chapter 2

BUYING YOUR EQUIPMENT

WHAT IS CB

Citizens Band (CB.) can be used by anyone, requiring a shortrange communication system, whether for private or business use. A person can talk about what ever they like over the air (as long as it is not an illegal or obscene activity.) It should be borne in mind that it is not a secure system so anyone can listen to, or join in with, an ongoing conversation. Popular belief is that CB cannot be used for business, this is wrong. On the contrary, it can be used for ones' profession, as long as the service is not used for advertising. CB radio is not only fun; Citizens Band is a very useful tool for anyone; on the road, at home or even at your business. Truckers and motorists use CB to keep things together on the road. It will warn them of changing road conditions, accidents and many other driving hazards stretching miles ahead, keeping them one jump in front. Many caravan clubs now use CB radio so that when there are 3 or 4 in convoy they can keep in touch and not lose one another in heavy traffic or when one has to stop in an emergency or to take on fuel.

With a set installed at home and another in your vehicle, it's a convenient way of keeping in constant contact in the event of that important 'phone call or message, as well as having personal communications with your "better half". CB radio also offers a means of keeping in reach of your business. Farmers, taxi drivers, driving schools and building contractors are just a few of the type of businesses using the CB radio system. There are many questions you will want to ask and so I will cover as many as possible. One of the most common (and the most difficult to answer) is "What sort of distance will I get?". The answer to this is dependent on many factors. Typical range on flat terrain between two mobile stations is about 5 to 6 miles, 8 miles between mobile and base and 15 miles between base to base. Handheld to handheld you can expect around 2 miles.

Where there are tall buildings, hills and trees the range is greatly reduced, particularly between two mobiles or if the base station is set in a valley, also if the system is being used by a great many people; in some cases the range will be reduced to under a mile.

Remember, it's the antenna that does most of the work. So spend a few pounds less on the CB. and use that on a better aerial.

MOBILE RADIOS

The first confusing part about choosing a radio (CB rig) is the fact that in the UK there are two legal frequencies on 27 MHz. The first will have the marking CB27/81. This frequency arrived when CB was first made legal in this country in November 1981 and the U.K. was flooded with various types of radios which could be bought from literally any corner shop. The second system will be marked PR27GB. This was made legal in the UK in December 1986 to come in line with European countries, and is known as C.E.P.T. or "mid-band" as some people call it; in some people's opinion, this system should have been adopted from the beginning.

It appears at this stage that the two systems will run side by side for some years to come. The one problem that I can see is that the older CB27/81 radios will become scarcer in the coming years as many manufacturers will stop production because the only country they are being produced for is the U.K.

CHOOSING A RIG

It will be found that any legal CB. that is bought will have 40 channels and a power out-put

of 4 watts (the maximum allowed). The front panel will have all kinds of fancy dials, knobs and switches on it, so lets take a look at what they do.

CHANNEL SELECTOR

The 40 channels will be in the form of a rotary channel selector so that any one of the 40 channels can be selected at will. Some of the later model radios have 'Up' and 'Down' buttons which step the channels up or down when pressed or flick through quicker when held down: the choice is yours.

VOLUME ON/OFF

This one really is self explanatory, it turns the CB set on and it adjusts the loudness of what the user is listening to.

SQUELCH

This is used to eliminate unwanted background noise and signals. It is usually set on the threshold where the noise is just eliminated. The further past the threshold, the stronger the station has to be in order for it to be received. Some of the weaker signals will be missed with the squelch set high. Any really strong signal will open the squelch and be received.

RF GAIN KNOB

This controls the reception sensitivity of the set. When turned down the set is less sensitive and when increased the set is more sensitive to more weaker signals. reducing the gain helps prevent overloading when talking to someone who is very close. Some sets instead of having a rotary control have a two position flick switch, see local-distant .

MIC GAIN

This control adjusts the sensitivity of the microphone amplifier circuit and controls the audio from the microphone that is used to modulate the signal. Ask somebody for a rig check to tell you at what point your signal starts to distort or break-up as the mic gain is increased.

METER

Most meters have two scales, one for transmitting and one for receiving. The first lets the operator know how strong the transmit signal is, the second informs the user how strong a signal is being picked up. There are quite a few radios around which have LED indicator lights; these replaced the conventional meter, but act in the same way as any moving coil meter. These meters cannot be 100% accurate and different rigs will give slightly different readings for signals of the same strength.

NOISE BLANKER (NB)

This switch cuts down on static, motor and ignition noise. It helps to filter out unwanted noise interference keeping it from getting into the receiver.

LOCAL - DISTANT (LOC/DIS)

This in fact an antenuator. In the distant position (the normal position used) it gives the full range and sensitivity of the receiver. This control is very similar to an RF gain, but in a switch form.

CHANNEL 9 SWITCH

This switch allows instant switching to channel 9 emergency frequency, from any other channel. When the channel 9 priority switch is used, the regular channel selector is disengaged.

P.A.

The P.A. control allows the user to turn the CB set into a public address system. Most CB's have a 3.5 jack socket on the rear panel that will take a P.A. speaker. The P.A. speaker is usually fitted under the bonnet of a vehicle.

LOW POWER SWITCH

(4 watt - 0.4 watt)

When this function is used it limits the transmit power of a set to 1/10th of it's normal power (0.4 watt). This switch will only normally be found on 27/81 model CB 's because it was a licence requirement when this system was made legal. As it is no longer a requirement, the newer C.E.P.T. radios rarely have them fitted.

ANTENNA WARNING INDICATOR (AWI)

This is a very useful function, but it is not found on many radios. If there is a fault with the antenna system, this light will illuminate when you transmit, to let the operator know that there is something wrong with the SWR.

MICROPHONE

One of the other devices the radio will have is a hand-held microphone, this will have a bar or button on the case. This is known as a PTT switch, meaning Press To Talk. It has to be pressed to talk to someone and released to receive. Unlike a telephone you cannot talk and listen at the same time, so all the time you are pressing the switch and talking a reply will not be heard. It must be understood that while communication is indeed two way, it can only be one way at a time.

SPEAKER

Most modern CB's will also come complete with a built-in speaker; this is usually found on the base of the unit, so attention must be paid when installing it into a vehicle so that the sound from the speaker can be heard satisfactorily.

POWER LEAD

Most CB's are designed to be fitted in negative-earthed vehicles, although a few will work in both positive or negative-earth vehicles. Other points to remember are the location, (there's no point in putting it somewhere where the controls are out of reach), or where the microphone lead will get caught up.

RX/TX INDICATORS

These light emitting diodes light up to show if you are receiving (RX) or transmitting (TX).

radio frequencies

PR 27 GB

1	26.965
2	26.975
3	26.985
4	27.005
5	27.015
6	27.025
7	27.035
8	27.055
9	27.065
10	27.075
11	27.085
12	27.105
13	27.115
14	27.125
15	27.135
16	27.155
17	27.165
18	27.175
19	27.185
20	27.205
21	27.215
22	27.225
23	27.255
24	27.235
25	27.245
26	27.265
27	27.275
28	27.285
29	27.295
30	27.305
31	27.315
32	27.325
33	27.335
34	27.345
35	27.355
36	27.365
37	27.375
38	27.385
39	27.395
40	27.405

CB 27/81

1	27.60125
2	27.61125
3	27.62125
4	27.63125
5	27.64125
6	27.65125
7	27.66125
8	27.67125
9	27.68125
10	27.69125
11	27.70125
12	27.71125
13	27.72125
14	27.73125
15	27.74125
16	27.75125
17	27.76125
18	27.77125
19	27.78125
20	27.79125
21	27.80125
22	27.81125
23	27.82125
24	27.83125
25	27.84125
26	27.85125
27	27.86125
28	27.87125
29	27.88125
30	27.89125
31	27.90125
32	27.91125
33	27.92125
34	27.93125
35	27.94125
36	27.95125
37	27.96125
38	27.97125
39	27.98125
40	27.99125

CB 934/81

1	934.0125
2	934.0625
3	934.1125
4	934.1625
5	934.2125
6	934.2625
7	934.3125
8	934.3625
9	934.4125
10	934.4625
11	934.5125
12	934.5625
13	934.6125
14	934.6625
15	934.7125
16	934.7625
17	934.8125
18	934.8625
19	934.9125
20	934.9625

Chapter 3

CITIZEN BAND RADIO FREQUENCIES

There are three sets of frequencies allocated to Citizens Band Radio (CB) in the UK. Only equipment that meets the appropriate Performance Specification or antenna restriction may be used. Listed next is what you can and cannot use together with some information on ancillary equipment. Note that the use of all other equipment is prohibited.

Apart from the 3 systems operating at the moment there will be a fourth promised for late 1996. Due to great pressure from licence holders and user groups, there will be an 80 channel radio comprising of the MPT 1382 and MPT 1333 rigs in one set. This system will only be legal for use in the United Kingdom.

HOW CAN I RECOGNISE LEGAL CB EQUIPMENT?

All CB radios bear a mark to show that they comply with the DTI Specification. These are :

MPT 1333
26.965 MHz - 27.405 MHz
ETS 300 135 (CEPT)

CEPT PR 27 GB

CB 934/81

MPT 1321
934.0125 MHz
- 934.9625 MHz
934 CB/81

PR 27/94

MPT 1382
27.60125 MHz - 27.99125 MHz
(PR 27/94)

CB 27/81

Pre: January 1996
(MPT 1320)

Can only be bought secondhand (CB 27/81)

WHAT IS LEGAL CB?

"Legal CB" refers to CB equipment that meets the specification laid down by The Department of Trade and Industry (DTI) for the frequencies used. A legal CB Station will in addition use an antenna (aerial) that will also conform with the restrictions.

CAN I USE RADIOS DESIGNED FOR MORE THAN ONE CB SERVICE ?

No, you will need a separate CB radio for each of the services you intend to use. The use of combined equipment is illegal.

CAN I USE CONVERTED EQUIPMENT

No, you may not use equipment that has been altered in any way. The use of Amplitude Modulation (AM) or Single Sideband (SSB) is prohibited. FM/PM has been shown to cause less interference in a domestic environment than AM/SSB and has been chosen as the European standard modulation for CB.

WHAT IS THE MAXIMUM POWER ALLOWED ?

The maximum RF power output allowed is 4 watts for 27 MHz equipment (MPT 1382/1333 and ETS 300 135), and 8 watts for 934 MHz equipment (MPT 1321).

WHICH FREQUENCIES MAY BE USED

The channels available for CB are shown in CITIZENS BAND RADIO FREQUENCIES shown earlier in this book.

CAN I USE ANCILLARY EQUIPMENT ?

The use of power microphones, echo boxes and speech processors is not illegal but neither is it recommended. In the case of equipment manufactured to the technical requirements of ETS 300 135, the microphone is subject to type- approval and hence must not be replaced with a different type of microphone. Locking push-to-talk microphones are not allowed.

One piece of equipment that is now legal is a "chat-back unit"; this is felt to be of particular benefit to the blind. It allows the use of a voice simulator to inform users which channel they are operating on. You may use an antenna receive pre-amp for 934 MHz operation (but not 27 MHz) although you may use a mechanical antenna switch. You may also use meters to set up a station. The use of power amplifiers (burners, boots etc.) is strictly prohibited.

CAN I USE A VSWR POWER METER ?

Yes, you can use a Voltage Standing Wave Ratio (VSWR) meter and an output power meter to set up a station, but they must be removed before the station is used as they may cause interference if left in line.

FACT-FILE

Schedule 2 of the WT (Citizens Band and Amateur Apparatus) (Various Provisions) Order 1988 provides for equipment approved in certain countries and marked with the appropriate country symbol, to be used in the UK on the condition that it has also been type-approved by the UK.

Legal CB equipment conforming to MPT 1333 or ETS 300 135 which has been approved in a country as listed below, will be marked as:
CEPT PR 27 + the symbol indicating original country where type-approval occurred.
e.g.. equipment type-approved in France will be marked: CEPT PR 27 F.
A full list of countries and symbols appear below:

Albania	AL
Austria	A
Belgium	B
Bosnia and Herzegovina	BH
Bulgaria	BG
Croatia	HR
Cyprus	CY
Czechoslovakia	TCH
Denmark	DK
Estonia	EST
Finland	SF
France	F
Germany	D
Greece	GR
Hungary	H
Iceland	IS
Irish Republic	IRL
Italy	I I
Liechtenstein	FL
Lithuania	LT
Luxembourg	L
Malta	M
Moldova	MLD
Monaco	MC
Netherlands	NL
Norway	N
Poland	PL
Portugal	P
Romania	RO
San Marino	RSM
Slovakia	SK
Spain	E
Sweden	S
Switzerland	CH
Turkey	TR
Great Britain and Northern Ireland	GB
Vatican City	SCV

Chapter 4

INSTALLATION

INSTALLING YOUR RADIO

. . . .When setting about installing a CB in a vehicle, there are several primary things to remember. Where to mount it is one of the first questions that comes to mind. This is often dependent on the driver's preference. Here are some of the practical things to keep in mind. Keep it within easy reach and clearly visible, (you don't want to end up in the rear of another car because you were trying to change channel). It should be kept out of the way of your gear lever and hand-brake. Make sure that any wires or microphone lead won't get tangled up in the steering wheel or brake and accelerator pedals. In trucks it is often possible to mount the unit securely above the driver's head, well out of the way, until required.

Heat can damage your CB, so don't mount it in the close vicinity of a heater. Also, it should be installed in a place where it won't get kicked or take a chunk out of your knee every time that you get into your car. So it usually turns out that the best place is under the dash-board close to the driver.

Another favourite position is between the two seats. Wherever the rig is to be placed, before finally drilling holes, plug in the microphone and make sure that the trailing microphone lead is not likely to get in the way -this point is often forgotten. Most CBers clip their mic to the side of the rig, but separate mic clips (magnetic, adhesive) are available to hold the mic almost anywhere, more convenient than having to fumble about under the dash. All rigs are supplied with fixing kits and it's important to follow the instructions carefully.

It is often a temptation to make light work of the job and avoid drilling for some of the fixing screws, or make other short cuts. One other major point, BE CAREFUL not to drill into anything behind the dash, such as electrical wires or heater hoses.

CB has got where it is today because of it's great possibilities for communications whilst driving down the motorway or through a large city. Since most people start out with a CB in a vehicle. Let's run through some of the aspects on how to install one in a car. Installation of a CB set in a motor vehicle or a truck is as simple as fitting a car radio. By reading the instructions supplied with the set and following some basic rules, results can be achieved without too much trouble. . . . .

CONNECTING THE SUPPLY

The next thing is to get some DC (12 volt) supply to it. The first priority is to check that the vehicle is NEGATIVE earth, (there are not many positive earthed cars about any more) this can be easily done by lifting the bonnet and checking the negative (BLACK) lead from the battery. If it is bolted to the chassis of the vehicle, it is negative earthed.

For temporary use of a CB rig in a car, the simplest possible power connection has got to be the cigar lighter. Simply remove the cigar lighter insert and put it in a safe place for future use. Cigar lighter plugs can be purchased from any car accessory or CB shop. These insert into the lighter socket and stay captive. Most have to be fused with an in-line fuse holder using the CB power lead - others have fuses in the plug itself. In any event the cigar lighter socket is usually fused at around 10 amp, which is reassurance but represents quite a large fault current and in-line fusing of 2 amp is recommended.

Having checked that the car is negative earth, then the outer case of the cigar lighter plug will be negative, with the centre pin, therefore, obviously positive. Connecting the cigar lighter plug to the CB is best done with a twin cable of 10 amp rating.

Although the set will not take that amount of current, a 10 amp cable is sufficiently robust not to be damaged if accidentally caught, in the glove box door for example. In many cases the CB manufacturer will supply a power plug which is non-reversible for the back of the set.

This can be soldered directly to the cigar lighter plug. For a more permanent installation the best place to pick up the supply is directly from the battery. Running it through the ignition system can result in unwanted ignition interference.

The CB radio needs two power connections, one RED and one BLACK. The RED wire will go to the POSITIVE (+) terminal and the NEGATIVE, BLACK, is best made directly to the car body at a point near to where the CB set will be mounted. It is usually possible to put this under one of the mounting bolts. If the positive wire has to pass through metalwork, a 5 mm hole should be drilled and lined with a rubber grommet to protect

CHECK IT OUT

Never replace the fuse with anything bigger than a 3 amp. If anything goes wrong, the fuse will blow, not the CB radio!

the cable. This is very important as vibration will wear through the insulation in no time at all.

Remember to use the power lead that came supplied with your rig, this will have an in-line 2 amp fuse. Never replace the fuse with anything bigger than a 3 amp. If anything goes wrong, the fuse will blow, not the CB radio!
If you are installing a CB into a truck it will be no doubt be 24 volt. So this will have to be reduced down to 12 volt by the use of a "dropper"; these are available from any good CB shop. Make sure that this is at least a 5 amp otherwise it won't be adequate for your CB.

ANTI-THEFT DEVICES

A CB radio set can be installed in such a way that it can be removed easily from the vehicle.
I recommend the fitting of an accessory called a slide mount, (sometimes called a sledge). CB's are small, expensive and light and are attractive bait for the small-time car thief; they do get stolen regularly, so it makes a lot of sense to remove the set when the car is left unattended.

The slide mount is worth every penny. Slide mounts are generally very reliable but the sliding contacts can tarnish and become less effective in time, requiring periodic cleaning with an electrical switch cleaner. One section of the mount is attached to the CB radio bracket (or bolted directly to the set) and it's wires are connected to the contacts on the slide mount. The other section is installed into the vehicle and its contacts in turn are wired into the vehicle's power supply and sometimes into the antenna line, although I do not recommend connecting the antenna system because this can sometimes result in a high SWR reading.

The two sections slide together and make contact. Removal is accomplished by unlocking the mount with a small key (supplied) and pulling the set towards you, disengaging it from the in-car section.

Chapter 5

MOBILE ANTENNAS

OK., you have now chosen your rig, and installed it, it is now time to make it work. Every CB has to have some sort of an antenna (aerial, twig, whip) connected to it, allowing the transceiver to transmit and receive properly.

IT HAS TO BE A CB ANTENNA. YOU CANNOT USE YOUR NORMAL CAR RADIO AERIAL.

For mobile use the antenna is usually a vertical whip. These can be between 2 and 6 feet long. The "whip" will have a loading coil of some description in its length. This coil, which is usually tightly-wound wire, can be added at the bottom (base loaded), centre (centre loaded) or top (top loaded), but it can be anywhere.

A base loaded antenna will normally be made of stainless steel and are generally very durable, but there are many helically wound antennas and some (DV27 for example) that have copper wire embedded in fibreglass. A point to remember here is that in general the longer the antenna is the better it will perform, the only short antenna I have found to work satisfactorily is the SIRIO T2. Also the higher on the vehicle you mount it the better you will get out and receive. As the antenna does most of the work the other factors to remember are;

(1) when mounting a base loaded antenna, keep in mind that the base radiates most of your signal, so it should be placed in a position clear of surrounding metal surfaces, on the roof is the most common place;
(2) buy an antenna that has a common fitting at the base so that if the "whip" does get broken it will be easy to replace;
(3) make sure that it can be taken off when the vehicle is unattended, they do get stolen quite often. A quick-release adaptor is a good investment.

CO-PHASED ANTENNAS

Co-phased or twin truckers as they are so often called, only really work when they are mounted at least 9 feet apart, as on the wing mirrors of a truck or caravan. Two identical antennas are used with a special cable known as a co- phased harness. This harness is nothing more than two equal lengths of 75 ohm coax cable terminating at one connector.

The two antennas combined together provide a directional signal back and forwards; this is why they are usually installed on trucks because they usually running up and down motor ways and need to know what is in front or behind them.

If the two antennas cannot be mounted 9 feet apart, then forget it, better results are achieved by using a single antenna on the roof or gutter of a vehicle.

WHERE TO PUT THE ANTENNA

The next step is the installation of a CB antenna on the vehicle. When mounting an antenna it is as well to remember that the metal body of the car itself becomes part of the antenna system, and where it is placed will affect the radiation pattern of the signal. As in all antennas, the higher it is placed the better it will receive and transmit, so the obvious place is in the centre of the roof, but this is not always practical and so a compromise must be made between size, performance and height. There are generally four types of mobile antenna mounts;

(1) drill a hole in the body of the vehicle
(2) magnetic mounted
(3) gutter mount
(4) hatch-back or boot-lip mounting, or if it is a truck
(5) mirror mounting.

The best and most permanent method of mounting a CB antenna involves drilling a hole in the body of the vehicle. This is ideal if you already have a convenient hole somewhere. Unless you own a van, roof mounting is very difficult in a car as it is virtually impossible for an amateur to remove the roof lining. Care must be exercised when drilling the roof, and an antenna mount that does not require access underneath. The cable is usually taken up through the trim covering the front window post. This is really a job for the professionals. Mounting on the boot lid or wing is relatively simple and involves carefully drilling the correct size hole in the car body. For those without experience, it is a risky job, not to be recommended.

Remember when fitting it to the body of a vehicle do not let the antenna run along side large areas of metal-work, as this will affect getting the SWR down to an acceptable level.

The gutter mount, as it suggests, fits onto the rain gutter of your car. The cable is usually fed through a nearby door seal, but a better idea is to run the cable along the rain gutter and into the boot. Earthing is achieved by grub-screws which penetrate the paintwork. The hatch or lip-mount will fit onto your boot or hatch-back.

These are the most common ways of mounting antennas, because of the work involved. They simply grip onto the edge of the boot lid. The cable is taped to the underside of the lid and a loop made where the lid hinges. Earthing is made with two small set screws which penetrate the paintwork under the lid. Feeder from the boot can be run the length of the vehicle

underneath the carpet. There are usually holes in the back of the car body adjoining the boot, behind the rear seat. If not a hole has to be drilled.

Magnetic mounts are a very convenient way of mounting an antenna, and are available for most types of antenna. Top and centre-loaded antennas do tend to wrench the magnet rather when they get a sway on. Mag mounts are by far the most popular CB antenna mount at the present time, because it dispenses with the tricky business of drilling holes in the body of the car. Generally mag mounts will not cling satisfactorily unless the whole of the magnet ring is in contact with metal. It is a good idea to clean grit and dust off the paintwork before fixing a mag mount to reduce any possibility of the paintwork being scratched should the magnet start to slide at all. The coax cable has to enter the car, usually through the boot lid or door seal. Mag mounts are easily removed to prevent theft or provide for easy garaging. Beware, they can tend to scratch your vehicle or inhibit the natural fading of the paint-work in time. Needless to say, a mag mount is of no use on a fibreglass bodied vehicle.

Mirror mounts are largely used on trucks and lorries that have brackets made of tubular steel to hold the mirrors, these are a popular mount for truckers in the USA. Do not try putting twin antennas (co-phased) on a car, this will be a waste of time. They can only work successfully if you can get them at least 9 feet apart, (5) as on the wing-mirrors of a truck.

ANTENNA REGULATIONS

Only antenna whose radiating element (excluding ground plane but including casing, coils and associated circuitry) is less than 6.95 m long and 55 mm in diameter are legal.

Dimensions of antennas and loading coils to be within a cylinder of the size shown.

Only single, vertical omnidirectional monopole antennas may be used. The use of Yagi or beam antennas is prohibited.

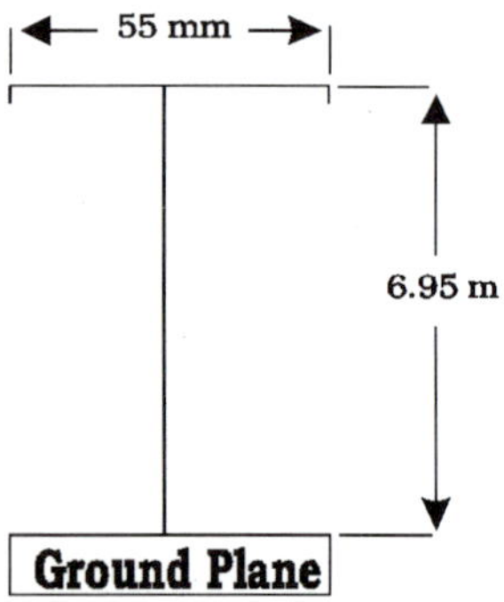

ARE THERE ANY HEIGHT RESTRICTIONS FOR CB ANTENNAS?

No, apart from local planning restrictions and a requirement that within 1 km of airfields the overall height of the antenna plus mast must be less than 15 m.

ANTENNAS FOR 934 MHz CB.

Any antenna with not more than 12 elements may be used. Each element may not exceed 17 cm in length.

ATTENUATION CHART

LOSSES IN DB **SOURCE**

1/2 dB . . . . .for every connector in line.
1 dB . . . . .for every SWR meter, switch or other add-on device.
2.5 dB . . . . .for every 100 feet of RG58 coax between radio and antenna.
1 dB . . . . .for every 100 feet of RG8 or RG213 coax between radio and antenna.

Chapter 6

BASE STATIONS

I suppose there are just as many base stations as there are mobiles about these days. CB need not be confined solely to mobile operation and a great deal of pleasure can be derived from contacting mobiles and other home stations from the comfort of an armchair.

There is also no reason why a mobile CB transceiver cannot be used at home as a base station. A mobile unit on a slide mount can easily be removed from the car and do double duty as a home base.
Using CB at home is very popular by numerous people, who are house-bound, for example, housewives, invalids and the older generation and at the other end of the scale, the younger generation not yet old enough to have a car.

Quite a few of the truckers rely on home-base users to relay traffic reports and road conditions. A great many CB users have a base and a mobile set-up so that of an evening they can still keep in touch with their "good buddies". (think of the money saved on 'phone calls). Setting up a home base is almost as easy as setting up a mobile unit.
The added advantage of having an antenna on the roof of a house means that you will get a great deal more distance than you will with a CB unit in the car.

INSTALLING A HOME BASE

You'll notice that a base station is a lot larger than a mobile set and that it runs straight from a 13 amp mains socket. This is because the power supply is built into the unit and it usually has many more features then a mobile rig e.g.; head-phone sockets, built-in SWR meters etc.; Some even have clocks! Several manufacturers supply dedicated home base stations that are housed in a cabinet suitable for use on a desk top.

When fitting a standard 13 amp mains plug to a mains powered base station, a fuse rating no more than 3 amp is required. The originally fitted 13 amp fuse should be replaced by a 3 amp fuse. When the transceiver is set out, the mains lead should be kept out of the vicinity of the antenna coax. This will reduce the possibility of 27 MHz pickup into the mains.

POWER SUPPLY UNITS

The other alternative to a genuine base station and a more economical way of doing things, is to have a mobile rig on either a DC power supply or a 12 volt car battery. A power source of 13.8 volt at around 2 amp is required to power the CB rig. Several manufacturers produce power supply units (PSU's) which plug into the household mains and have the correct 13.8 volt output.

As long as this is more than the current taken by the CB there will be no problem. As with the mains powered home base, a 3 amp fuse in the 13 amp plug top is required to protect the equipment. This plugs straight into a 13 amp mains socket and your red and black lead from the CB connect onto the two terminals on the DC supply. Red goes to the positive (+) and Black to the negative (-). Make sure that there is a 2 amp fuse in-line on the Red (+) lead.

USING A 12 VOLT BATTERY

It is possible to use a 12 volt car battery instead of a DC power supply. (see fig 1a). This is not as convenient as a DC supply as this will have to charged up about once a week.

Care should be taken as battery acid can and will eat away your clothes, carpets and lots of other things. It is advisable to keep it in some sort of plastic container and well out of reach of small children.

One handy thing about a car battery is that if you suffer a power-cut you are still "on the air" while the rest are in the dark.

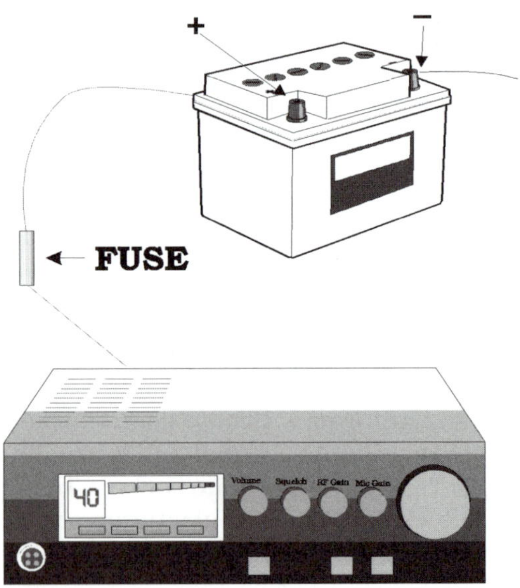

Chapter 7

BASE STATION ANTENNAS

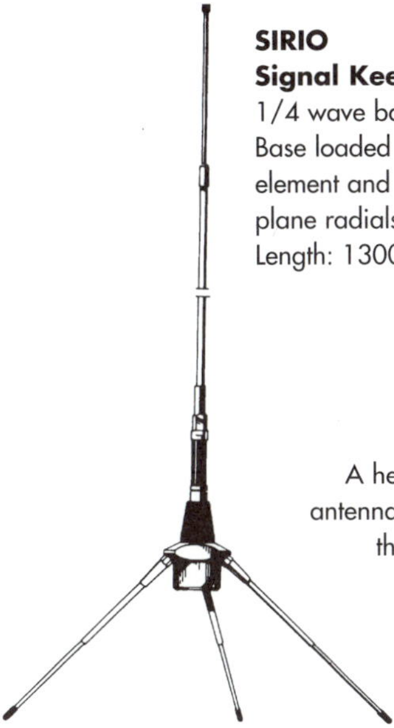

SIRIO
Signal Keeper 27
1/4 wave base antenna. Base loaded radiating element and 3 ground plane radials.
Length: 1300mm

SHAKESPEARE
Big Stick
A heavy duty fibreglass antenna that will withstand the worst of weather.
Length: 18′

SIRIO
GPS 1/2 Wave
Made of non-corrosive aluminium tubes and has jointing sleeves of polythene for perfect waterproofing.
It has a transparent coil cover and anti-static cage.
Length: 5.5 metres

Section 3a of the Citizens Band Radio Licence Schedule states that 27 MHz apparatus with provision for connection to an external antenna shall only be connected to an antenna which does not exceed 6.95 metres in length. The CB Band is, and is often referred to as 11 metres and so a full wave antenna for CB should be 11 metres in length but this of course is impractical and so CBers use 1/2 wave antennas which are 18 ft long.

A normal 1/4 wave antenna should have no dB (decibel) gain but as the antenna gets bigger so will the gain. A 1/2 wave antenna has a gain of around 3dB (see fig 2). and when you get up to the large Quad antennas the gain figure is around 11dB.

There are basically three types of base antenna. These are 1/4 wave, 1/2 wave and 5/8 wave, (although there are a few 3/4 wave antennas around) These will measure approximately 9, 18, and 22 feet in length. Quite a number of the base antennas will have radials fitted, these are rods protruding from the base at 90 degrees or more from the main vertical element. All of the above mentioned antennas are what is known as omni-directional, i.e. they radiate (receive and transmit) in all directions.

In general the theory is "the bigger the antenna the better", and the higher the antenna is mounted the better range of communication you will get. Most 1/2 and 5/8 wave antennas are designed to be mounted on a steel or aluminium pole ranging from 1 to 2.5 inches in diameter. These poles will normally be mounted onto wall brackets (known as T and K brackets) secured with a couple of U bolts.

There are also various chimney lashing kits available to enable securing the pole to a chimney-stack. Make sure that any brick-work you that are fixing to is sound and that a good quality Rawl-Bolt is used, otherwise the first sign of any strong winds you'll have to start all over again and replace the antenna.

DB GAIN TABLE

DB GAIN	MULTIPLY BY	DB GAIN	MULTIPLY BY
0	1.0	7	5.0
1	1.2	8	6.3
2	1.6	9	8.0
3	2.0	10	10.0
(where most people notice a difference)		11	12.6
4	2.5	12	15.8
5	3.0	13	20.0
6	4.0		

Space is often a big problem and so many CB stations use a mobile antenna at home. This will have to be mounted on some sort of ground-plane to simulate a vehicle. The ideal size would be a sheet of metal at least 4 feet by 4 feet square. When space is a problem the most common indoor antenna used is a DV27 mounted on a biscuit tin. This will work moderately well, but it is a compromise.

Chapter 8

HAND-HELD CB

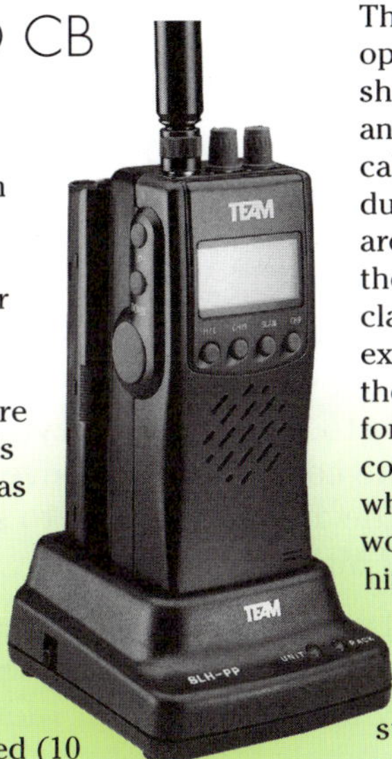

Portable CB's are transceivers with an internal battery supply and built-in antenna. Portable or Hand-helds can go anywhere and have countless uses. There are many hand-helds (or "walkie talkies" as they are sometimes called) around, some on 49 MHz and the more popular, on 27 MHz. The 49 MHz variety are very low-powered (10 milliwatt) and therefore will give no great distance. Because of this, 49 MHz hand-helds do not require a CB licence.

27 MHz hand-helds will come with two types of power outputs, either 2 watt or 4 watt, the latter of the two being the better. A 4 watt 40 channel is the best hand-held to purchase for a number of reasons; They are usually good for communications up to 3 or 4 miles, perhaps further if connected to an external antenna.

A 4 watt hand-held is powered by either 8 alkaline batteries (8 x 1.5 volt = 12 volt) or 10 rechargeable ni-cads (10 x 1.2 volt = 12 volt). If alkaline batteries are to be used then 2 dummy batteries have to be installed; these are usually supplied with the handheld. Almost all CB hand-helds come with a 1.5 metre telescopic antenna; it should be remembered that before transmitting on a hand-held the telescopic antenna must be extended to its full length otherwise the SWR will be too high when the antenna is in the down position. There are some optional add-on, short, rubber helical antennas (often called "rubber duck") around that are about 10" long, these can be slid or clamped on to the existing telescopic, they are very useful for short-range communications when the 5' antenna would be a hindrance. Auxiliary jack inputs on many hand-helds connect to such optional equipment as: battery chargers, 12 volt DC leads, earphones,external microphones and antennas.

A 40 channel hand-held usually comes with all the features of a good mobile rig but with the added advantage of being able to turn off the LED display to conserve battery life. There is usually a switch to reduce power to some 400 milliwatts for short range work, which also considerably lengthens battery time.

Two popular items from **TEAM**

BELOW

MINICOM FM

A handheld at the leading edge of current technology. It delivers a full 4W output and boasts a comprehensive range of features.

LEFT

PORTY PHONE

This radio can be used with the optional "drop in" charger making this really practical to use all day long!

Chapter 9

S.W.R.

SWR, which is short for Standing Wave Ratio, always seems to strike fear in many people, because it is the portion that seems to be cloaked in mystery. It really is very simple and to get the best performance it should always be adjusted.

To adjust the SWR, an SWR meter and a short patch lead will be needed, this will cost just under ten pounds, but worth every penny, because the operator can check the SWR regularly to make sure everything is working properly and in the long term save the cost of an expensive repair. OK. What we are doing here is to adjust the antennas length to suit the radio frequency being used (in this case 27 MHz).

If this is not done and the set is used with a bad SWR, some of the reflected power is bounced back into the coax and into the radio itself. This reflected power can reduce efficiency and damage your radio's power transistors in time.

ADJUSTING THE SWR

Right, we are now ready to SWR your antenna! Connect one end of the patch lead to the SWR meter socket marked XMTR, connect the other end to your rig ANT socket. Connect your antenna lead to the SWR socket marked ANT. Now tune your rig to channel 20. Put the SWR switch to the forward (FWD) position. Now press the microphone button and adjust the knob or slider on the SWR meter until the meter reads SET (the end of the scale).

Now keeping the microphone keyed flip the switch to the reflect (REF) position and note the reading on the SWR scale then release the microphone key. If this reading is less than 1.5 your antenna is tuned and it is not worth the trouble trying to get it any lower. If after checking you find that your SWR is 2 or more, you will have to tune your antenna.

Your first job is to examine your coax and connectors. If these are OK. next check that the antenna mount is making good contact with the body work of the vehicle, these can be the major cause of bad SWR's. Assuming that these are all satisfactory your next step will be to adjust the antenna. First take a reading on channel 1 and channel 40. Which had the higher SWR?. This will tell you if it has to be longer or shorter. If the SWR was lower on channel 1, the "whip" has to go down. Move it about 1/4" at a time.

Most antenna's have a grub-screw which can be loosened to adjust the "whip". If the "whip" won't go down any further into the coil you will probably have to cut some off. Cut no more than 1/4" at a time, be careful not to cut too much off. Remember to re-set the SWR meter after every fresh reading. If the SWR is higher on channel 1, then the reverse applies, the "whip" will have to go up in length.

When carrying out SWR adjustments, be sure to keep the car away from large objects such as an up-and-over garage door, make sure nobody is standing close to the antenna and that the doors, boot and bonnet are closed as all these can contribute towards a false reading. Basically it's the same procedure to SWR your base antenna, but because of the length you will find this a much easier job to carry out. The same rule applies here, check out readings on channels 1 and 40. This will tell you whether the antenna has to be longer or shorter.

One final point, make sure the SWR meter you are using is functioning OK. and that the patch lead is in good condition.

SWR PROBLEMS

A persistently high SWR is likely to be the result of a damaged coax feeder, or incorrectly fitted plugs and at this stage an ohm meter will help to clear up the problem. If it looks like the antenna is at fault, a 50 ohm dummy load connected in its place should enable an SWR of 1.2:1 to be achieved easily unless the feeder is damaged or the CB is incorrectly set up internally.

If this is the case then repairs and adjustments will have to be carried out by a properly equipped CB service department. If all of the antenna system seems to be working correctly and it will not tune to an acceptable SWR, it is possible that an antenna tuner (matcher) will improve matters.
A matcher is connected between the rig and the antenna and must be left permanently in line. (see fig 4a) The matcher is inserted in the coax feeder line at the CB end using a short patch lead. The SWR meter is still connected to the CB set before the matcher to enable tuning of the matcher to be accomplished. Set the rig on channel 20, and adjusting each of the matcher controls in turn will show which way each one should be turned to reduce the SWR and at some point both controls will have achieved maximum reduction in the SWR reading.

It is as well to check the band edges (channels 1 and 40) and in some cases a compromise may be necessary to achieve a moderate SWR over the band rather than optimum results over just a few channels. The matcher stays in-line permanently, but the SWR meter can be removed. It is a good idea to tape over

CHECK IT OUT

CAN'T GET THE SWR DOWN

If you cannot get the SWR down, here is a little tip. If the reading is lower on channel 1 than 40 you know that the antenna must be shorter, i.e.; go down. If it is higher on channel 1 than 40, it wants to be longer, i.e.; go up.

FACT-FILE

Under Section 4 (2) of the Citizens Band Licensing Schedule a meter to measure the transmitter output power or Voltage Standing Wave Ratio (VSWR) of the antenna may be TEMPORARILY inserted between the antenna socket of the transceiver and the antenna for the sole purpose of establishing a CB station or testing equipment providing that such meters are removed before a CB station is used. In other words, you cannot keep the SWR meter in-line during use.

the controls of the matcher to eliminate the possibility of unintentional adjustment which, if it goes unnoticed, apart from reducing the range of the transceiver, could cause damage to the rig with a high SWR.

Puncturing a myth (1): NOTHING you can do to the set, the cable or any other part of the system will or can affect the antenna's SWR. A "matcher" can fool the set into thinking the SWR is good, but an untuned aerial is an untuned aerial.

Puncturing a myth (2): The SWR cannot and will not vary no matter how much or how little power you use. Always remember to use the "FWD" or "SET" adjustment if and every time you change the power you are using. To those who say they CAN see a difference, remember that all SWR meters (at least those costing less than about 1500 pounds) are increasingly insensitive as the power drops, so that a reading of 1:1.2 measured at 400 mW can really mean 1:2.5, which is what you would measure at 100 watts or so.

CUTTING THE COAX WILL CHANGE MY SWR

False, false, totally false. there is NOTHING you can do at the CB end of the system which will change the SWR at the antenna. The problem can only be corrected at the antenna itself. And if you've heard somebody claim his SWR changed as he trimmed the coax cable, that was a definite indication of a problem at the

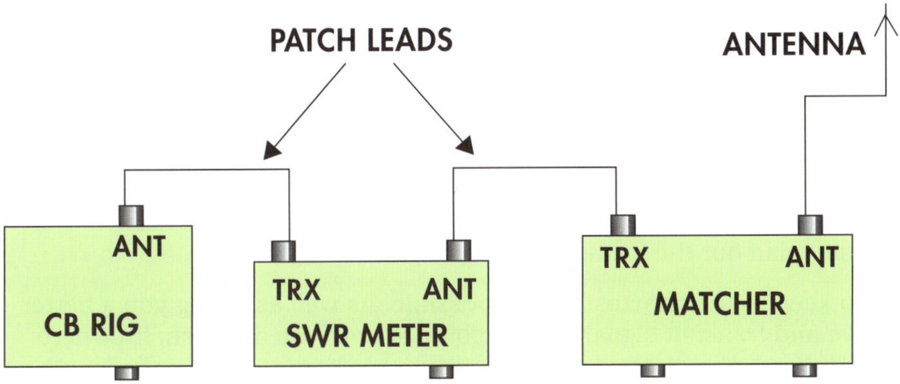

antenna end. What actually happened was that due to a faulty antenna, the coax itself had become a part of the overall radiating system, so cutting it affected the entire system. Only the antenna should affect the SWR reading.
The "impedance" of all CB radios is "50 ohms" by design; we can't change that. And our coax cable is also "50 ohms" by design, so we can't change that either. This leaves us with the weakest link, the antenna itself, which could be any "impedance" from a few "ohms" to a several thousand "ohms". But you are clever enough to realize that if your CB antenna could be "50 ohms" then BINGO! you have the perfect 1:1 SWR. So what you are actually doing when you adjust the antenna tip up and down is trying to make that antenna become "50 ohms impedance".

In very simple terms, SWR is simply the comparison of how much CB transmitter power is getting out into the air, where you want it, to how much is "reflected" back down the coax into the CB, where it doesn't belong. That's why your SWR meter is marked "FWD" meaning "forward power" and "REF" meaning "reflected power".

You may occasionally see something about "VSWR", meaning "voltage standing wave ratio". This is really what your meter is reading; i.e. forward and reflected voltage.

Chapter 10

ANTENNA CABLE

Antenna cable, or coax as it is known, can be bought from any good CB shop by the metre. Don't use TV cable, this is 75 ohms and it's the wrong impedance. You need 50 ohm cable for your CB. All CB antennas are made to use this type of coax cable. Before setting off to buy your cable, carefully plan out the route you are going to run it.

Try to keep to the shortest length possible. As well as giving you a better receive and transmit signal, it isn't going to cost you as much. It pays to add 5%, just to be on the safe side, and do keep your coax in one length, don't start to join it.

You will lose 1 dB for every connector in the coax line. If your cable is going to be under 50 feet you will be safe using a good quality RG 58, this is around 40p a metre. If your cable run is going to be over 50 feet, you would be better using RG 8 or RG 213. You will not get as much signal loss through this type of coax. For example; if you transmit 4 watts through 5 feet of RG 58 cable, about 80% of the power reaches the antenna; with RG 213 90% reaches it. And likewise, if you run 100 feet of RG 58 you get about 65% at the antenna, and with RG 213 about 80%.

A good quality cable such as RG 213 is around 1 pound per metre. Don't forget that the same figures apply to any signal you are trying to receive. It is an "Old Wives" tale that cutting sections of your coax will reduce the SWR. This may change your SWR slightly, but it won't help you get out any better. What it means is that you are SWRing the cable, not the antenna. The same tale goes for keeping your cable in 9 feet lengths, this will make no difference to the SWR whatsoever.

Making sure that you have soldered the PL 259 and that it isn't "shorting-out" (this is done with the use of a multimeter or a bulb and battery), tape the cable to the pole at the antenna end, don't let it hang free. Leave a short loop. This will reduce any stress on the PL 259 should the antenna sway in any high winds. And likewise, where it enters the house, leaving a small loop will stop rain running down it and into your home.

If your SWR is high, make sure the antenna is the right length and has been properly assembled, also check out the connections at each end. Any water getting into the cable will corrode the braid and ruin it. The only cure for this is complete replacement, so make sure everything outside is waterproofed, especially the plugs and connections.

CONNECTIONS

Fitting a PL 259 plug to RG 58, RG 8 or RG 213 cable is quite a straightforward task with the aid of a good hot soldering iron.

CHECK IT OUT

Check out your cable for "shorts" with a multimeter once you have fitted the PL 259 plugs to each end. NOT when the cable is connected to the antenna.
As most antennas are DC grounded this will give a false reading.

ASSEMBLY INSTRUCTIONS FOR PL259

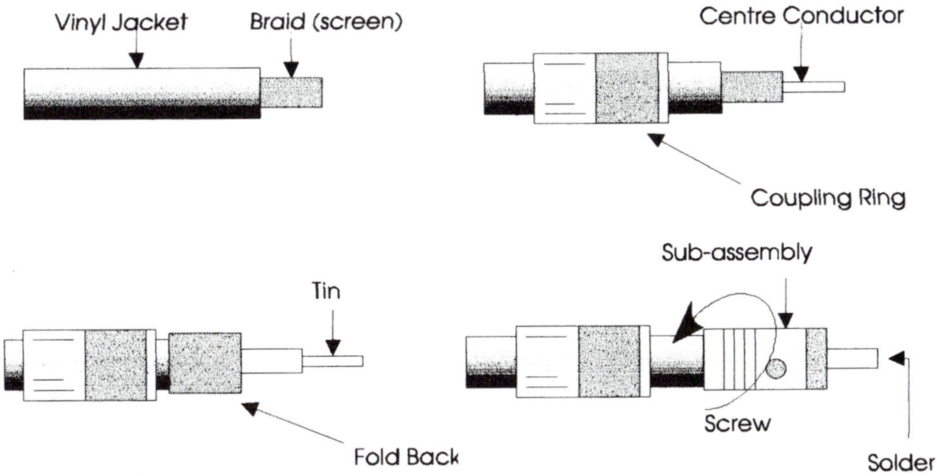

Chapter 11

USING YOUR CB RADIO

OK. the CB radio has been installed in either a vehicle or at home. The antenna has been erected on the house or installed on the car. And (I hope) you have adjusted the SWR to a satisfactory level. Most people on Citizens Band radio use a "handle", it's a type of CB code name. If it is thought about, there would be hundreds of Johns or Mikes out there but hardly any of "The Spirit" for example. Now it's your turn to select one. When this is done, try and choose a handle that isn't too long and one that's easy to pronounce.

Perhaps in the beginning you will want to try a few until one is found that you get on with. There was a breaker in my area who called himself "Ayatolla", and many times I heard others calling him "Iron Collar".

Now the moment we've all been waiting for, key-up the microphone and talk to someone. WAIT....... at first just listen for a few hours, take a spin through the channels and monitor some of the conversations. In the beginning it takes a little time for the ears to get adjusted to the radio and at first everything sounds a bit fuzzy because of the interference. I have heard many a person who is listening to CB for the very first time say "What did he say?" when to me it was quite obvious because I'm used to the audio from a CB radio.

In most areas, the channel a conversation is started on (known as the breaking channel) is channel 19 or channel 14. You will have to listen in your town to find out which one it is. The breaking channel is a kind of meeting point, once that contact has been made with someone, move on to one of the quieter channels, first making sure that it's not being used by someone else.

There is a good chance that the very first time you key-up the microphone and try and talk to someone, you will experience "mic fright". In other words, you will get nervous and dry-up. This is completely normal and has happened to many a new breaker. Just remember that if you're getting embarrassed and going scarlet, it is being done in private and no one else can see you. If at first you don't succeed, try again.

When first starting to talk to someone, just talk in normal language. Don't worry about the CB Lingo or the Ten Codes (covered later in this book), these will come to you in due course. There are many sensible breakers out there who will know that you are new "on air" and will only be too willing to help you.

CB is a fun thing, but on a more serious note, it is best to be wary about giving out an address or telephone number because unless you have met them, nothing is known about the person being spoken to (or who else is listening for that matter). However exciting a breaker may sound on air, you don't want the 'phone ringing at 3 in the morning or a complete stranger knocking on the door at all hours. If you do wish to meet a person that is spoken to, make it a neutral place like the local pub or club until you get to know them better.

FACT-FILE

Any equipment used must be marked with either "CEPT PR27 GB" or "PR 27 GB" to indicate that they comply with DTI Specification MPT 1333 (i.e. operate on 26.965MHz -27.405MHz). On 11 June 1991 the final version of ETS 300 135 was published. This will eventually replace DTI Specification MPT 1333. It is envisaged that there will be a two year paralleled running period during which equipment may be type-approved to either Specification. Following this all new equipment will be type-approved to ETS 300 135. Such equipment will be marked with "CEPT PR 27 GB" as before. All equipment type-approved to MPT 1333 may be used for the lifetime of that equipment.

USING YOUR CB ABROAD

CAN I USE MY CB ABROAD?

COUNTRIES THAT HAVE IMPLEMENTED CEPT T/R 20-02 AND 20-07

- **Austria • Belgium • Cyprus** *
- **Denmark** (equipment marked "CEPT PR 27 GB" only)
- **Finland • France • Germany**
- **Luxembourg**
- **Netherlands**
- **Norway** (equipment marked "CEPT PR 27 GB" only)
- **Portugal** (equipment marked "CEPT PR 27 GB" only)
- **Sweden • Vatican City**

* Awaiting confirmation that Cyprus will accept equipment marked with either "PR 27 GB" or "CEPT PR 27 GB

Yes, but only CEPT CB i.e. 26.965 - 27.405 MHz and with certain restrictions. Countries that have implemented CEPT (European Conference of Postal and Tele-communications Administrations) Recommendations T/R 20-01 and T/R 20-07 will allow UK CB licensees operate under the terms of their UK licence. A full list of such countries is listed. Similarly CB licensees from these countries may operate in the UK under the term of their licence.

Licensees should ensure that they carry with them a copy of their up-to-date licence and a copy of the licence Terms and Conditions Booklet. Schedule 2 of the WT (Citizens Band and Amateur Apparatus) (Various Provisions) Order 1988 provides for equipment approved in certain countries and marked with the appropriate country symbol, to be used in the UK on the condition that it has also been type-approved by the UK.

When operating abroad, licensees should first and foremost, comply with the conditions of the appropriate licence of the country that they are visiting. In addition licensees should comply with the conditions of their UK licence unless to comply would breach the conditions of the country visited. Any breach of the licence conditions of the country visited could be regarded as a breach of a licensee's UK licence as well as being an offence in the country visited.

Chapter 12

INTERFERENCE AND ABUSE

Legal FM Citizen's Band equipment when it is operated within the terms of the licence is less likely to cause interference than illegal equipment using AM or SSB modulation.

INTERFERENCE TO TV AND RADIO

Advice on reducing or eliminating interference to domestic radio and television sets is contained in a comprehensive booklet "How to Improve Television and Radio Reception" published by the Radio communications Agency.

INTERFERENCE TO DOMESTIC ELECTRICAL EQUIPMENT

Where equipment that has not been designed to pick up radio signals is adversely affected by transmissions, the fault lies in the equipment itself. This may include equipment such as home computers, hi-fi's and other domestic electrical items. Advice on curing the problem may be obtained from the equipment manufacturer or possibly from a local radio or hi-fi dealer. The Radio Investigation Service (RIS) cannot investigate complaints of interference to any equipment that is not designed to pick up radio signals.

INTERFERENCE TO OTHER CB EQUIPMENT

The Citizen's Band radio service was accommodated in the already crowded radio spectrum only on the basis that it would not be "protected" - that is, whilst CB equipment must be maintained and used so that it does not cause any undue interference to other radio services, users of CB cannot be protected from interference from other radio services. The radio amateur is in the same position. The RIS acts against operators of illegal equipment where their activities are causing much interference to the legal CB service.

WHAT CAN BE DONE ABOUT INTERFERENCE FROM FOREIGN STATIONS

Interference from foreign, often Italian, stations is not limited to the CB service, though it is a particular problem at 27 MHz during some stages of the sunspot cycle. This is simply one of the features of the 27 MHz band. Enforcement of licence conditions in foreign countries is the responsibility of their own administrations.

WHO IS RESPONSIBLE FOR CURING INTERFERENCE FROM A LEGAL CB SET

If CB equipment is found or known to be deficient and causing interference, the CB operator is responsible under the terms of the CB licence to see that appropriate corrective action is taken. The radio amateur is in the same position. Advice on curing interference problems may be obtained from equipment manufacturer or possibly from a local CB dealer, or from a good CB handbook. Moving the set, cable or antenna a few feet may cure the problem.

WHAT IS BEING DONE ABOUT ABUSE ON CB

The RIS has the task of monitoring radio frequencies to check for licence abuse, including "channel blocking", the playing of music and using obscene language. The Wireless Telegraphy (Content of Transmission) Regulations 1988 make it a specific offence to use obscene, offensive or menacing language in transmissions.
The Police can directly follow up such offences. Insofar as resources allow, the RIS takes action against people who use CB radio unlawfully and cause a serious nuisance to other users. Threats are a matter for the Police.

In taking action against those who abuse the CB service, the RIS has the right to seize and seek forfeiture of any equipment that is connected to the station or any other item which is considered to be of evidence of the offence.

If you wish to make a complaint about licence abuse, including "channel blocking", the playing of music and use of obscene language and have specific evidence relating to the complaint you should write, giving full details, to your local Radio Investigation Service District Office.

CODE OF PRACTICE

The operating conditions of the CB service has been deliberately made simple with few restrictions. Although it has always been made clear that no-one has preferential rights at any time or place or on any channel, a recommended voluntary Code of Practice has been agreed by CB User Groups and is printed in the CB licence conditions booklet. Among other things the Code asks that priority be given to calls for help, and in particular recommends that channel 9 be left clear for emergencies and assistance only. All operators are asked to follow the Code and so promote good operating practice within the CB service.

Chapter 13

EMERGENCY AND CHANNEL 9

The CB Radio Code of Practice contained in the CB licence conditions booklet (obtainable free from all Post Offices with the Licence Application Form) recommends that all CB radio operators follow operating convention and use channel 9 only for emergencies and assistance.

IS CHANNEL 9 A "LEGAL" EMERGENCY CHANNEL

The Code of Practice recommends that the channel be left clear for emergencies but this does not constitute a legal requirement and is not a licence condition.

WHY ISN'T CHANNEL 9 PROTECTED BY LAW FROM ABUSE

Volunteers do valuable work by giving up their time to monitor channel 9 for emergency calls and their frustration when the channel is misused is understandable. But legal protection for the channel is not an easy remedy for channel abuse because it would have to involve effective checking and enforcement action. The cost of providing resources on a large enough scale to do this would be difficult to justify and would put up significantly the cost of the CB service.

DO I HAVE TO REGISTER WITH THE AGENCY TO BECOME A MONITOR

No, any group or individual licensee can monitor channel 9, or indeed any other channel. No permission is needed from the Agency, and the Agency does not maintain a register of monitors.

SHOULD I REGISTER WITH THE EMERGENCY SERVICES

This is not necessary but you may like to contact the emergency services to let them know you are there (some like to keep a list of known CB monitors in their area) and to get any advice they may wish to give you about the passing of emergency calls.

It is also important to get in touch with the local police if you wish to help in incidents such as searches for lost children. Sometimes the emergency services can be hindered rather than helped when people turn up on the scene of an accident or search and it is therefore very important to make sure that your efforts are properly directed.

ARE MONITORS EXEMPTED FROM ANY LICENCE RESTRICTIONS

No, a CB monitor, like any other user, must comply with the conditions of the CB licence at all times. For example, it is an offence to use equipment which is not permitted in the licence and there are no exceptions to this. Some operators are under the impression that large antennas can be installed for use in an emergency or for reception only, but this is not the case. Similarly, the use of power levels greater than 4 Watts (at 27 MHz) is an offence.

In short, channel 9 monitoring is a useful part of CB, but it must be carried out within the normal rules of the CB service.

DO I HAVE TO BE A MEMBER OF A TEAM OR MONITORING ORGANISATION

Many channel 9 monitors do belong to a local or national monitoring group and this brings the benefits of collective experience and good training. It is not necessary to join such groups, however, and you are quite free to monitor on your own if you wish to do so.

DO THE EMERGENCY SERVICES MONITOR CHANNEL 9

Generally no, some monitoring may be done locally by services such as police traffic controls but this is not usually on a regular basis. The emergency services certainly do not have sufficient resources to undertake monitoring on anything but a small and selective scale. CB is no alternative to the maritime emergency service, for example.

Chapter 14

SKIP AND BLEEDOVER

SKIP

This refers to at least two different forms of atmospheric propagation which allow much further contacts than normal. Under "flat" conditions, the usual distance given earlier (up to about 20 miles) will apply; this is about as far as the 'ground wave' will allow.

However, on occasions (usually around the middle of the day, more rarely in the evening) good signals, sometimes very strong, will appear from stations about 400 - 800 miles away. These may be, for example, from Scotland to Hampshire or even England to Germany or Italy. These signals arrive by sporadic-E (Es for short). They are reflected by what is known as the "E-layer". They can be quite short-lived, maybe just a few minutes, or they can go on working for several hours. They arrive at quite a high angle, so quarter- or half-wave antennas are best for them.

The real long distance version is "F-layer" skip. It is rare to hear anyone less than 1000 miles away when this is running, but it can (and on occasions does) reach right round the planet!

Puncturing a myth (3) : When conditions were really good some years ago, some of the strongest signals from the States and elsewhere on AM seemed to have an 'echoey' quality. Thus arose the legend that echo was a good thing to have for DXing. What was really happening was that the main signal was coming direct across the Atlantic; the same signal was also coming the long way round via the Pacific, India and Central Europe and arriving a moment later because it had further to go. It could even go round again and give a third "echo".

Food for thought: If echo improved your DXing chances, all the Radio Amateurs, Armed Forces, commercial operators etc. would be using it. They don't!

The F2 layer is very high up in the ionosphere, with a lower angle of reflection (5/8 wave antennas) and allows contacts with Africa, North and South America and even Australia. The best chances of finding these contacts is on the CEPT band, which is the same as the American FCC frequencies and are used virtually anywhere that CB is permitted. Early evening is usually the best time for "F" propagation, tailing off to Central America at around midnight.

F2 working has to follow the "sunspot" 11-year cycle; it gets good in the summer for about 2 years in every 11, and as this is completed in late 1995 we are just about at the down peak of one.

BLEEDOVER

The term "bleedover" covers a multitude of different forms of interference. What is not often realised is that most of them are the fault of the receiver, not the transmitter. The only important ways in which a transmitter can give bleedover is if it is over modulating and spreading it into the next channels on either side (he will probably sound rough, quiet and muffled on his own channel) or if it is faulty and/or very badly aligned ("tweaked") and is actually generating a spread of signals as well as his own.

By far the most common causes of bleedover arise in the receiver. If the needle drops back and the breaker you are listening to disappears or becomes very faint, it is because a very strong signal on another channel is blanketing the RF amplifier. (Cure: Use the RF Attenuation or Gain control). If you hear the voice of the interfering station across a lot of channels, maybe even all of them, he is completely overloading the first mixer and his signal is blasting its way through by sheer force.

The cure is the same, or get a rig with a wider 'Dynamic range'. It may be suggested that a "bleedover crystal" (a crystal filter) is a good idea to fit to your rig; it can be and will certainly reduce 'adjacent channel interference' but it cannot fix these two.

"Adjacent channel" interference can arise from the filters in the IF section being too wide but can also be created within the PLL circuit that generates the local oscillator "channel" frequency for the receiver, for technical design reasons which are too deep for this short book to cover. A crystal filter can give a distinct improvement to some rigs, but not all, and some sets are simply not designed to take one. "Co-channel" interference is self evident from its name; someone else using the channel at the same time as you means you can only use short range, to someone even stronger than the unwanted station.

This form of interference often occurs with 'skip' running, particularly from Italy, hence the dreaded term "wall to wall spaghetti".

Different rigs are more or less prone than others to all of these forms of interference; some of it can be reduced, some can't.
The best you can do is make sure that the rig is properly installed and set up, and that it is correctly adjusted. DO NOT attempt to tune or "tweak" your rig unless you know EXACTLY what you are doing and why, and don't let anyone else do it either unless they are a properly qualified radio technician with all the right equipment. This description does not fit the vast majority of "rig doctors" there are around - you have been warned!

Chapter 15

TELEVISION INTERFERENCE (TVI)

When the frequencies of 27 MHz and 934 MHz were allocated to the CB Radio service it was recognized that interference could be caused to other radio users in the UK. Although there is more likelihood of TVI with an AM CB set, it is also a problem with FM, and therefore it is worth making an effort to minimize or eliminate it. This will enable you to live at peace with your neighbours.

Before you start to use the set check out your own television set for interference. TVI is the bugbear of CB, and the thing which most annoys the rest of the non CB users - and even other CB users, who put up with it in good spirit, but don't like it just the same.

The slightest glimmer of interference on your neighbours television and you will get the blame, simply because they can see your CB antenna and know that you have a CB radio. If you can show the trouble is not coming from your direction they will calm down.
The design of the TV is usually at fault, and not always with the CB set. Some makes are more prone than others, and it is thought that if a television set has a remote-control it is more likely to suffer.

HOW TO STOP TVI

The first thing likely to cause TVI is the power of the transmission. The use of linear amplifiers to boost power is almost certain to cause television interference whatever steps you take, and the only solution is not to use them (which is what you are supposed to do) or to limit their use to when TV programmes have ended, or mobile use only.

The next thing that helps reduce the problem is having a low SWR, The better the SWR is the less chance there is of TVI.

The most successful cure for TVI is the installation of a high pass filter (HPF) connected to the back of the offending television. The HPF is inserted in the TV aerial lead and the fly lead from this is plugged into the TV set in place of the aerial. (see illustration on the right).

FACT-FILE

The Home Office tests found that 27 MHz transmissions were causing interference in more ways than one:
* In some cases they were directly affecting the colour circuit, which led to the colour disappearing.
* In others the CB signal was directly entering the video circuit of the TV. This led to the picture being affected or disappearing.
* In still others the sound circuits were affected, and noise heard from the TV speaker.

Fitting a low pass filter, (this fits between the rig and the antenna) which cuts out high - frequency harmonics, is also another cure. (see fig 6)

When installing your home base antenna, keep this as far away as possible from any television aerials and their cables and use a good quality coax cable yourself.

Interference is not limited to televisions, it can also affect hi-fi systems, radios and in extreme cases telephone apparatus.

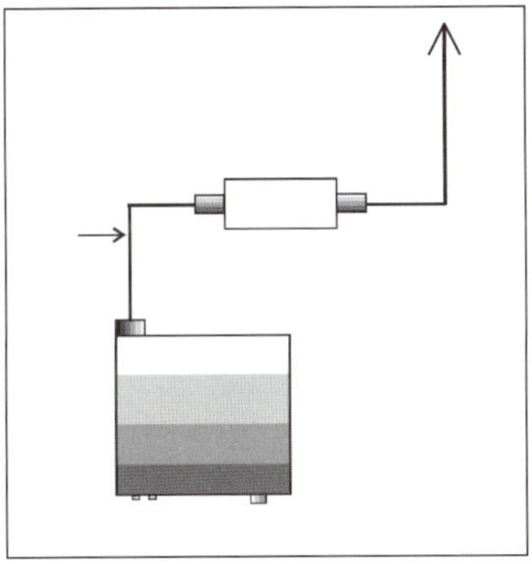

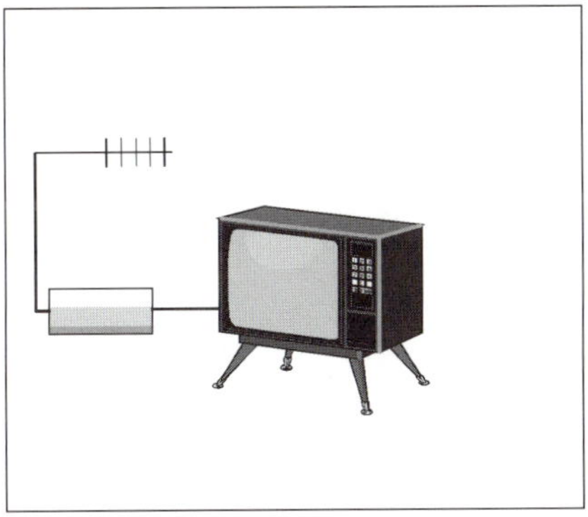

Remember that CB can cause interference, (which is why the government at first refused to have 27 MHz at all) and the main responsibility lies with the CB user in reducing TVI, and the amount there is will be the biggest factor in determining whether CB is happily accepted by the general public.

Chapter 16

HOW A CB RADIO WORKS

A CB Radio, despite the fact that it appears to be a sophisticated contraption, is really a very simple machine that works on the same basic principle as wireless transmission pioneered by Guglielmo Marconi in 1895. Nonetheless, most individuals do not understand exactly what happens when they hit the switch.

Right, let's have a quick trip through a CB Radio. When a set is receiving, a radio signal arrives through the antenna, slides down the antenna cable and into a radio frequency amplifier, where it is amplified and sent to the mixer. At the same time as it enters the mixer, a signal, generated within the set by an oscillator, is also fed into the mixer.
This new signal has a frequency that predetermines the selection of a particular channel. When a channel is selected, an oscillator circuit (the PLL) generates the proper channel frequency to tell the receiver what frequency to pick out and send to the speaker.

The selected signal then goes through a filter and enters the intermediate frequency section. Now called an intermediate frequency, usually at 10.965 MHz, the signal is amplified and then put through the detector, which extracts the audio portion of the signal from the carrier portion. The signal is then passed through a noise limiter and sent to the audio amp, where it is given a final boost before being put through the loudspeaker.

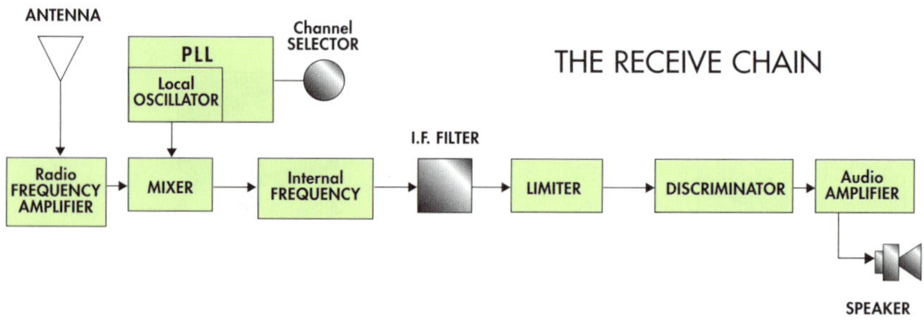

CB Radio sets are transceivers. They are capable of transmitting and receiving radio signals. When your set transmits, this is basically what happens. By pressing the PTT (push-to-talk) button on your microphone a relay is activated and your set becomes capable of transmitting. Voices are converted in the microphone into audio signals and fed into an audio amplifier. Here they are amplified and sent to the modulator, which is a detector in reverse. Instead of removing a carrier, the modulator adds a carrier signal to the audio signal, now the signal is fed into a power amp. When the PTT button is pressed, the oscillator generates a weak signal. The frequency of this signal is determined by the channel selection. This weak frequency signal is then put into a buffer amp. From here the frequency signal is put into a power amp, along the coaxial antenna cable, and out to the antenna. Quite simple really isn't it?

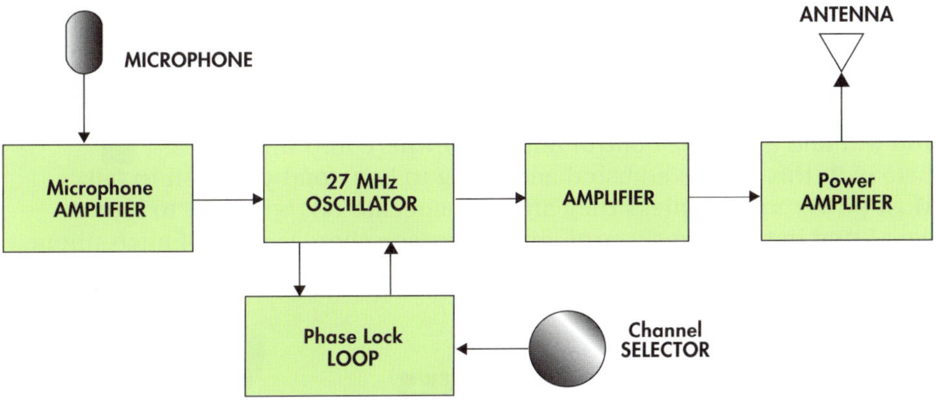

THE TRANSMIT CHAIN

CAPTURE EFFECT

With FM there is no Heterodyne or squeal when two stations key-up together, as there is on AM. This is because of the unique "capture effect" of FM. An FM receiver will capture the strongest signal on channel, any other weaker signals on channel will be overridden.

In some aspects this limits the range of an FM signal where CB is very active, such as a large city. If you are talking to a CB station and a stronger one starts to transmit on the same channel, go and make a cup of tea, you will have to wait until the strong station stops transmitting before you can carry on talking to your weak station.

Because CB stations only hear the stations close to them, two, three or even four sets of conversations can be going on at the same time in different parts of the city without interfering with each other.

Chapter 17

ACCESSORIES

POWER MICROPHONES

Power mics come in all shapes and sizes. What you will find is that they will not increase the deviation of an FM transceiver. The mic that is supplied with your rig is selected to give all the necessary deviation. Adding a power mic will possibly improve your quality of audio, but if it is turned up too high it can cause your voice to distort. Any power mic has a built-in amplifier and for this reason a battery will have to be installed in it.
You will find a volume control built somewhere into the case of it, this can be adjusted according to how loud you wish to talk. Most power mic's (unless they are for a specific CB) will have to have a plug fitted onto the curly-cord, for this reason there is a list of microphone connections later in this book.
COST: £10 - £50

BASE STATION MICROPHONES

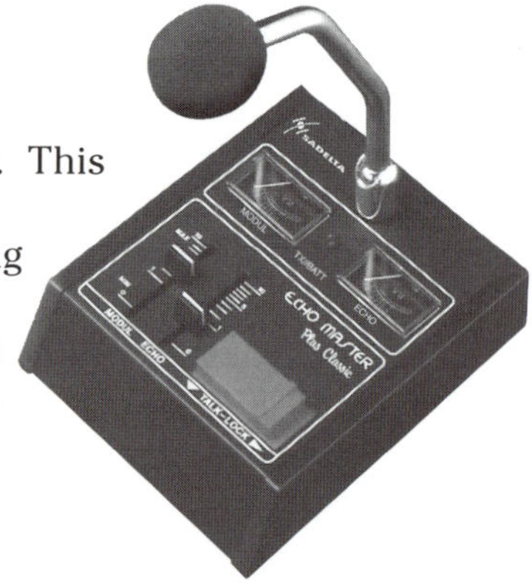

The ideal Base Station accessory is a Base mic. This is a larger "stand-up" version of the hand-held power mic but has the added facility of allowing the user to lock-down the mic key, giving a "hands free" operation. Base mic's also require a battery to be installed although some do run from 12 volt DC.
COST: £25 -£90

EXTERNAL SPEAKERS

As most CB radio's get installed under the dash of a vehicle, an extension speaker, plugged into the 3.5 jack at the rear of your CB radio, will get the sound up to where you need it and add an extra dimension to listening.
The use of an extension speaker can result in

louder reception and better fidelity. For example, the driver of a noisy lorry will find a definite increase in reception having a speaker up in front of him.
COST: £5 - £14

FREQUENCY COUNTERS

Frequency counters only really work in transmit mode and so you will only get a frequency reading when the microphone is keyed-up. When you key the mic the frequency counter will flash-up a digital readout of 5 or more digits depending on the model being used.
For example, if you were to transmit through the frequency counter on channel 19 on a C.E.P.T. version CB radio, the frequency 27.185 would be seen. There is a full frequency listing in the BUYING YOUR EQUIPMENT section of this book.
COST £40 UPWARDS

P.A. HORN

P.A. (standing for Public Address) horns can be installed to any CB that has a P.A. system on it. They are usually mounted under the bonnet of your vehicle facing the front.
These are a handy item for outdoor functions such as, fetes, sports events etc.
COST £8-£15

DUMMY LOAD

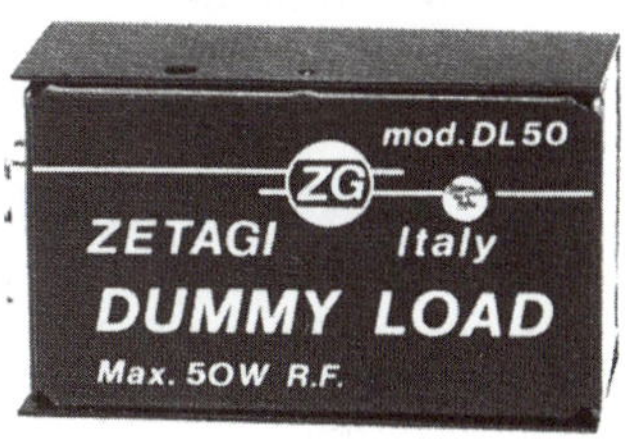

This device is a precise 50 ohm resistive load. With a dummy load you can carry out checks on your CB radio without taking up valuable airspace. The only way to get an exact power reading on your rig is to use a dummy load.
A dummy load has to be used in conjunction with an SWR / Power meter. **COST £15 UPWARDS**

MOBILE LINEAR AMPLIFIERS

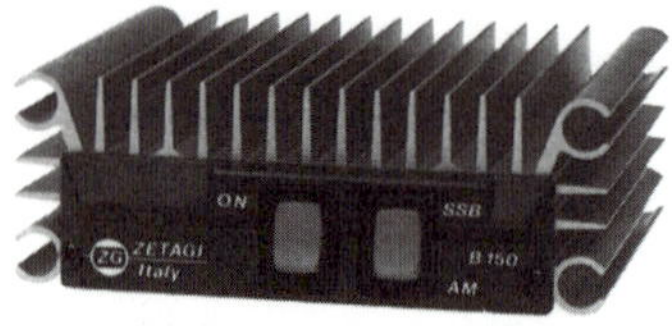

Although they are illegal, I am including Linear Amplifiers in the accessory section for it's completeness. Firstly, for those who don't know what a linear amp is, it is an add on device for increasing the output power of a CB radio from its original legal 4 watts up to in excess of 400 watts. To a large extent linear amplifiers are responsible for interference to TV and audio equipment, hence the legal limitation of 4 watts.

For mobile use they operate directly from your 12 volt battery, although there are some 24 volt models available for trucks. They range from 25 watts for local use, up to 400 watts which allows for very long-range communications. When they are installed in a vehicle it is advisable to connect them directly to the battery, via their own fuse. It is possible that if they are connected to the fuse-box the fuse will not be adequate.

COST £16 - £290

BASE LINEAR AMPLIFIERS

The higher powered valve Linears are the usual favourite for base station use. The mobile 12 volt units can be used, but this does get expensive because of the cost of a large Power Supply to run them. Valve linears must be tuned by means of two controls, PLATE and LOAD. These must be adjusted one at a time to give the maximum power output, THE TWO CONTROLS ARE NOT FOR VARYING POWER as many people think, this will only damage the unit. Most base linear amps have a meter to assist in tuning for the maximum power. Recently there have been quite a few Solid State linears appearing on the market, these will not need tuning as they are wide band and rely on transistors and not valves.

COST £100 - £500

PRE AMPLIFIERS

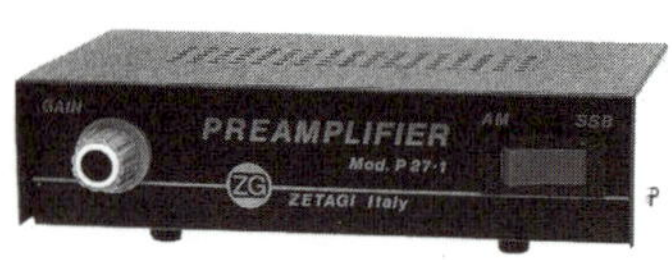

A receive Pre Amplifier works the opposite way to a Linear Amplifier; it will boost the receive signal by up to 25 dB (decibels). Quite a few pre amps now have a control so that you can adjust the receive gain. According to the Licensing Regulations, unless you are using 934 MHz equipment, a Pre Amplifier should not be connected in-line on 27 MHz CB radio equipment. **COST £20 -£35**

FACT-FILE

Under section 5 (i) of the Citizens Band Licence Regulations the use of linear amplifiers is illegal, the maximum power allowed is 4 watts.

SELECTIVE CALL UNIT

Sel Call units as they are known, are mostly used by persons using CB Radio for business use, which is perfectly legal in the UK When used with a CB radio a sel call unit will override the squelch control of another CB station being called who is also using selective call. The CB radio of both (or more) stations is muted until either one is selectively called by the other. At present there are not many CB radio's fitted with sel call, apart from the TEAM models.

COST £60

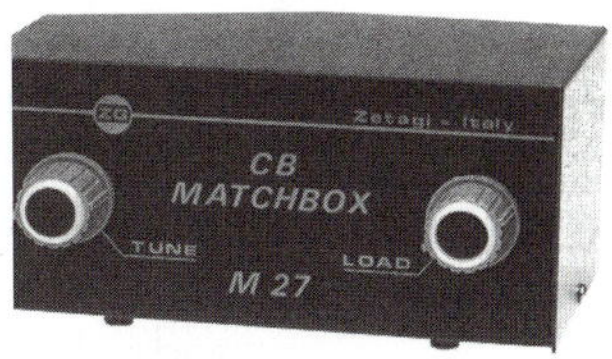

ANTENNA MATCHER

If you have a high SWR that you cannot get down in the conventional manner, then a useful accessory is an antenna matcher; this will prevent any high SWR from damaging your transceiver.

Antenna matchers have to be used in conjunction with an SWR meter, although some more expensive models have an SWR meter combined.

By adjusting the 2 controls on the matcher you will get the SWR down to a minimum. Antenna matchers do not improve the SWR at the antenna, they only "fool" the CB radio into thinking that it's OK. Your CB will perform much better if you can get the SWR down without the aid of a matcher. See also SWR PROBLEMS covered earlier in this book.

If you have an SWR reading of 1.5 or better, installing an antenna matcher will not improve the situation, in fact the opposite will occur. You will lose signal because of the insertion loss of the accessory in-line.

COST: £8 - £80

SLIDE MOUNTS

Because of the increase in theft of CB equipment a slide mount is a good investment. Slide mounts come in two sections, one half is connected to the CB and the other half is installed in your vehicle. The two halves of the mount slide into one another and make contact. Now when your car is left unattended the CB unit can be removed with ease and taken with you. **COST £8**

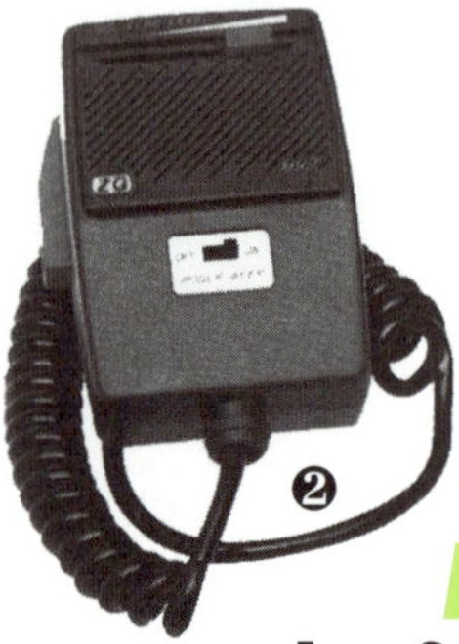
❷

❸

❹

MICROPHONES

1 SADELTA ME3 - A handheld echo mic, the electret insert provides the necessary "punch" in audio clarity.

2 ZETAGI M99 - Handheld echo/power mic with roger beep.

3 SADELTA 7 - A handheld mic which incorparates a good quality ceramic insert which provides top quality output.

❶

4 ALAN +3 - A pre-amplified mic with vu-meter, TX indicator, volume control, 9Vdc battery & 13.8Vdc external supply voltage.

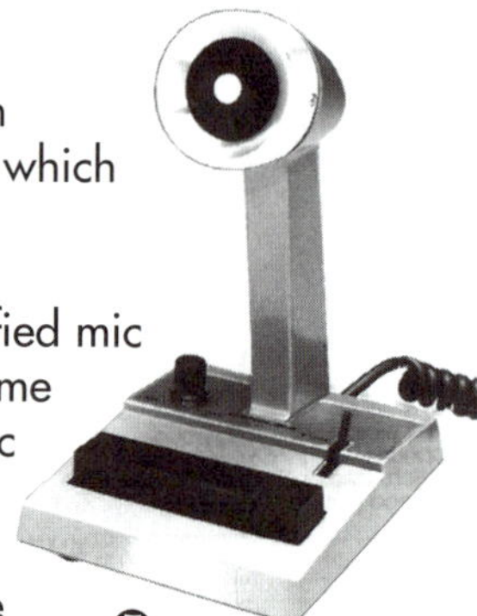
❺

5 TW232 - A reliable base microphone with built-in speech compressor that gives "full" modulation.

6 SADELTA ECHO MASTER PRO - A futuristically designed base echo/power mic. Three stage controls for delay, echo & modulation is achieved via 3 independant sliders whilst audio output is monitored via the modulation meter.

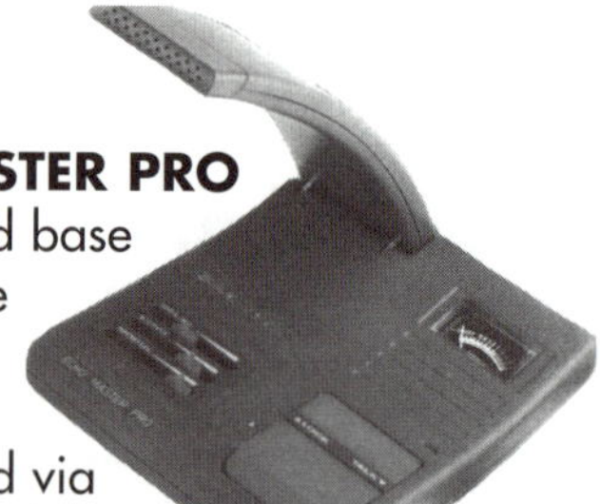
❻

Chapter 18

MICROPHONE CONNECTIONS

Because the microphone plays a major part in the operation of a CB radio, I have included just about every mic connection that I could find (legal FM sets). A malfunctioning CB is caused 80% of the time by a faulty microphone. One broken or disconnected wire can stop a CB from transmitting, receiving or modulating, it can stop all three if the common (shield) is at fault.

The quickest answer is to try another microphone that is known to be working. But sometimes this is not as simple as it seems. Just because your CB microphone has a 4 pin plug connection on it doesn't necessarily mean a friends will work on it, they are not all wired the same. For example, a Cybernet radio is wired completely the reverse from a Uniden although they both use a 4 pin microphone plug.

One other reason for including the microphone connections is that when going about buying a new replacement or power microphone for your CB it will be found that the majority of them do not come with a plug fitted. Your local CB shop will normally fit this for you, but if they cannot (and some can't) this can be done quite easily yourself with the aid of a good soldering iron.

PINOUT	MAKE & MODEL
4	Academy CB501
4	Academy CB502
1	Amstrad CB900
1	Alba CBM1
1	Alpha 4000
1	Amstrad CB901
2	Audioline 340
2	Audioline 341
Special	Audioline 342
4	Barracuda GT868
1	Barracuda HB940
1	Beta 1100 (Kernow)
12	Beta 2100 (Kernow)
12	Beta 3100 (Kernow)
Special	Binatone B/phone
1	Binatone Route 66
1	Binatone Speedway
1	Binatone 5 Star
1	Braemar PT40
4	Cheiza CB702
3	Cobra 21XFM
3	Colt 295
3	Commtron CB40F
4	Commtel GT858
4	Commtel GT868
1	Communicator NI440
1	Compact 40
1	Consam 1320
1	Cybernet Beta 1000
1	Cybernet Beta 2000
1	Cybernet Beta 3000
1	Cybernet Delta 1
5	DNT M40
1	Domico Convoy 1
1	Eurosonic GT868

PINOUT	MAKE & MODEL
1	Domico Convoy 1
2	Danita 640
7	Danita 440
1	Eurosonic GT868
1	Eurosonic ES404
1	Eurosonic Euro II
1	Falcon FCB1281
4	Fidelity CB1000FM
1	Fidelity CB2000FM
1	Fidelity CB2001FM
1	Ford Roadmaster 505
7	Grandstand Bluebird
7	Grandstand Gemini
1	Grandstand Hawk
1	Harrier CB
1	Harrier CBX
2	Harry Moss 325
1	Harvard 402 (H160)
11	Harvard H403 G/Buddy
1	Harvard 420M (H405)
1	Harvard H646
1	Icom ICB1050
1	Interceptor TC300
1	Jesan KT4004
1	Jesan KR200
1	Jesan KT7007
1	Jesan Pro 8000
1	Kernow Beta 1100
1	JWR M2
1	Kestral
4	Lake Manxman 850
4	Lake Manxman 950
5	LCL 2740
5	LCL Economy
1	Lancaster

PINOUT	MAKE & MODEL
5	LCL Enterprise
11	Legionairre
3	Lowe TX40
3	Maxcom 4E
3	Maxcom 6E
3	Maxcom 20E
3	Maxcom 21E
1	Maxcom 30E
3	Maxcom Appolo 16E
1	Midland Power Max
12	Maxon MX1000
12	Maxon MX2000
1	Midland 2001
1	Midland 3001
1	Midland 4001
3	Midland 77-099
3	Midland 77-104
13	Midland 77-805
1	Midland 77-095
1	Midland Power Max
1	Moonraker Major
1	Moonraker Minor
1	Murphy DS602
1	Mustang CB1000
1	Mustang CB2000
1	Mustang CB3000
3	Nentone
1	Pama GX19
1	Pama GX25
1	Pama GX 1000
1	Pama GX 2000
1	Planet 2000
5	Radiotechnic RT852
1	Radiomobile CB201
1	Reftec 934

PINOUT MAKE & MODEL

6	Realistic TRC2001
6	Realistic TRC2000
6	Realistic TRC2002
Special	Realistic Remote
1	Rotel RVC220
1	Rotel RVC220
1	Rotel RVC240
1	Rotel RVC230
1	Satcom Scan 40F
4	Sapphire X2000
1	Sapphire X4000
8	Shogun
3	Sirtel Searcher
1	SMC Oscar 1
1	SMC Oscar 2
1	Spinneytronic CB199
1	Sun 401
12	Team Eurocontrol
1	Team TRX404UK
1	Team TRX404
1	Team TSM404
12	Team TS290
3	Team TS1000
1	Transcom GBX4000
1	Uniace 400 (943)
2	Uniden 100
2	Uniden 200
2	Uniden 400 (cept)
2	Uniden PC404
2	Uniden PRO420E
2	Uniden PRO450E
1	Viper 88
1	York JCB861
1	York JCB863
1	York JCB867
7	Zodiac M144
7	Zodiac M244

MICROPHONE **PINOUTS**

PLAN 1
1 = Audio
2 = Screen
3 = Receive
4 = Transmit

PLAN 2
1 = Screen
2 = Audio
3 = Transmit
4 = Receive

PLAN 3
1 = Audio
2 = Transmit
3 = N/C
4 = Screen
5 = Receive

PLAN 4
1 = Receive
2 = Transmit
3 = Audio
4 = Screen

PLAN 5
1 = Audio
2 = N/C
3 = Transmit
4 = Screen
5 = N/C

PLAN 6
1 = Screen
2 = N/C
3 = Transmit
4 = Audio
5 = Receive

PLAN 7
1 = Audio
2 = Screen
3 = Transmit
4 = N/C
5 = Receive

PLAN 8
1 = Audio
2 = Screen
3 = Transmit
4 = Receive
5 = 12 Volt Feed

PLAN 9
1 = Screen
2 = Audio
3 = Transmit
4 = N/C

PLAN 10
1 = Screen
2 = Screen (linked)
3 = Transmit
4 = Audio

PLAN 11
1 = Audio
2 = Screen
3 = Transmit
4 = Receive

PLAN 12
1 = Audio
2 = Receive
3 = Transmit
4 = Down/Up
5 = Screen
6 = N/C

PLAN 13
1 = Transmit
2 = Receive
3 = N/C
4 = Screen
5 = Audio

4 PIN

SHOWN FROM SOLDERED SIDE

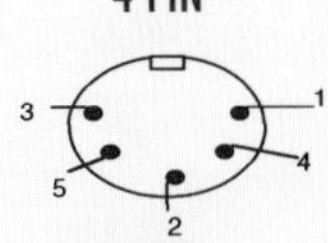

5 PIN DIN

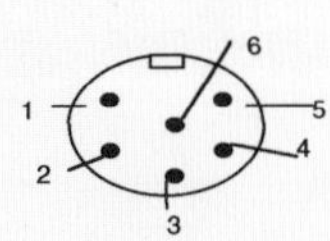

6 DIN

N/C = No Connection

MICROPHONE **COLOUR CODES**

BASE STATIONS

PINOUT MAKE & MODEL

PINOUT	MAKE & MODEL
9	Antron CB5097
2	Audioline 345
5	DNT B40FM
4	Fidelity 3000FM
9	Grandstand Base
1	Harrier HQ
1	Harvard H404
1	Harvard H407
10	Murphy CBH1500
6	Realistic TRC3000
1	Satcom Scan 4000
1	Team Euro 3100UK
2	Uniden 300
12	Team Euro 3004
2	Uniden Pro 620E
9	Wagner 506

Here are some of the more common microphone colour codes.

OLD TURNER 3 WIRE
shield............common
white ......mic or audio
black ....TX
red..........................RX

NEW TURNER 5 WIRE
red & shield..common
white......mic or audio
blue........................TX
black.....................RX
yellow ....not used

SADELTA (ALL)
shield ....common
white.......mic or audio
brown ....TX
green......................RX

K40
black & shield..common
white ......mic or audio
red ....TX
blue ....RX
yellow...........not used

ALTAI (ALL)
black & shield..common
red.........mic or audio
white ....TX
blue ....RX

CTE F10 & F16
shield ....common
blue ....mic or audio
white.......................TX
red ....RX

REALISTIC
shield............common
white ......mic or audio
red ....TX
black......................RX
blue.............not used

HEATHERLITE
shield ....common
white......mic or audio
red ....TX
black.....................RX

ZETAGI
shield ....common
yellow................audio
black.......................TX
red ....RX

Chapter 19

CB CODES

TEN CODES

I've listed most all of the 10 code for your convenience, but you will find that only about 15 - 20 of these are used in the UK.

10-0	Take care/caution
10-1	Poor reception
10-2	Good transmission
10-3	Stop transmitting
10-4	OK. Message received
10-5	Pass message on
10-6	Busy/Stand by
10-7	Out of service
10-8	In service
10-9	Repeat message
10-10	Transmission completed, standing by
10-11	Talking too fast
10-12	Visitors present
10-13	Advise weather/Road conditions
10-14	A party at..
10-15	Disturbance
10-17	Urgent business
10-16	Make a pickup at...
10-18	Anything for me
10-19	Return home
10-20	My location is...What is your location
10-21	Phone me
10-22	Report in person to...
10-23	Stand by
10-24	Assignment complete
10-25	Can you contact
10-26	Disregard message
10-27	Moving to channel...
10-28	Identify your station
10-29	Time up for contact
10-30	Illegal use of radio
10-31	Crime in progress
10-32	Radio check
10-33	Emergency
10-34	Trouble here, help!
10-35	Confidential info
10-36	Correct time
10-37	Breakdown at...
10-38	Ambulance needed
10-39	Your message delivered
10-41	Change to channel
10-42	Road accident
10-43	Traffic jam at...
10-44	Message for you
10-45	Anyone within range
10-46	Assist motorist
10-50	Break channel
10-53	Accident /Road blocked
10-60	Next message number
10-62	Can't copy use 'phone
10-63	Network directed to.......
10-64	Network clear
10-65	Awaiting next message
10-66	Cancel message
10-67	All units comply
10-69	Message received
10-70	Fire at....
10-71	Proceed with message
10-73	Speed trap at...
10-74	Negative/No
10-77	You are causing interference
10-81	Negative contact
10-82	Reserve hotel room for...
10-84	Reserve lodgings
10-85	My telephone number is...
10-88	My address is....
10-89	Advise telephone number of..
10-90	Radio repairs needed
10-91	I have/I am giving TVI
10-92	Talk closer to the mic
10-93	Adjust your transmitter
10-94	Check my frequency
10-95	Give me a long count
10-96	Transmit 5 sec. carrier
10-97	Subject is mental
10-98	Check test signal
10-99	Mission completed
10-100	Going to the toilet
10-200	Police needed at.....

PHONETIC ALPHABET

Of all the international codes used on CB the most basic is the Phonetic Alphabet. It is an easy way of getting a message over if the person at the other end cannot hear you clearly.

Alpha.........A	**November..N**
Bravo.........B	**Oscar ....O**
Charlie.......C	**Papa..........P**
Delta ....D	**Quebec ......Q**
Echo...........E	**Romeo ....R**
Foxtrot.......F	**Sierra.........S**
Golf............G	**Tango.........T**
Hotel..........H	**Uniform.....U**
India I	**Victor.........V**
Juliet ....J	**Whiskey.....W**
Kilo ....K	**X-ray ....X**
Lima..........L	**Yankee.......Y**
Mike...........M	**Zulu ....Z**

CHARLIE
OSCAR
DELTA
ECHO
SIERRA

R - S REPORT CODE

The R - S report code has been in use for a great many years by DXers. Besides exchanging QSL cards, DX operators usually swap R - S reports to give each other some idea of what signal they are sending and receiving.

R

1.......Unreadable
2.......Barely readable
3.......Readable with difficulty
4.......Readable with little difficulty
5.......Perfectly readable

S

1........Barely perceptible
2........Very weak signal
3........Weak signal
4........Fair signal
5........Fairly good signal
6........Good signal
7........Moderately strong signal
8........Strong signal
9........Extremely strong signal

he Q code

The Q code is an international language used worldwide by, radio amateurs, sideband operators and skip talkers.

QRA	What is the name of your station?
QRB	How far are you from my station?
QRC	Will you tell me my exact frequency?
QRD	Where are you going and Where are you from?
QRE	What is your estimated time of arrival?
QRF	Are you returning to......?
QRH	Does my frequency vary?
QRK	What is the readability of my signals?
QRL	Are you busy?
QRM	Manmade electrical interference?
QRN	Natural interference, Static.
QRO	Increase transmitting power.
QRP	Decrease transmitting power.
QRQ	Transmit faster.
QRS	Transmit more slowly.
QRT	Stop transmitting.
QRU	Anything for me?
QRV	Are you ready?
QRX	When will you call me again?
QRY	When is it my turn?
QRZ	Who is calling me?
QSA	What is the strength of my signals?
QSB	Are my signals fading?
QSK	Can I break in on your transmission.
QSL	Acknowledge receipt.
QSM	Repeat last message.
QSN	Did you hear me on channel...?
QSO	Communicate with.....
QSP	I will relay.
QSW	Do you wish to transmit on this channel?
QSX	Listen to.....on channel.....
QSY	Change to channel/frequency....
QSZ	Send each word or sentence twice.
QTE	What is my bearing from you?
QTH	What is your position?
QTI	What is your course?

QTJ What is your speed?
QTL What is your heading?
QTN Departure time.
QTR Correct time?
QTS Will you send your callsign for tuning purposes.
QTU Which hours is your station open?
QTV I will listen for you on channel....
QTX Listen for me until....
QUA Have you news of....
QUD Have you received the emergency signal sent by...
QUF Have you received the distress signal sent by....
QUM Is the emergency traffic ended?
QUO Shall I search for...?
QUR Have survivors been picked up?
QUS Have you sighted survivors or wreckage?
QUT Is the position marked?

ONE OF THE MAIN PASTIMES OF DXERS
is the sending and receiving of QSL cards

Chapter 20

TECHNICAL TERMS

If you are going to get involved in CB radio then you will obviously come across quite a few technical words. Most of them will be explained in this book, but I have put them all together to make life easier.

AC . . . . .alternating current (as in mains electricity).

AM . . . . .amplitude

MODULATION . . . . .Method of putting voice signals on to carrier wave.

AMPERE (amp) . . . . .unit of electrical current.

AMPLIFIER . . . . .device for increasing the level of any signal.

ANGLE MODULATION . . .another name for FM

ANGLE OF RADIATION . .the plane at which signals radiate from the antenna. A low angle is preferable.

ANTENNA . . . . .CB aerial. Radiates a signal from a transmitter and receives signals from other transmitters.

BAND . . . . .part of the frequency spectrum used for a particular purpose - used by CBers to refer to their 40 channels.

BANDWIDTH . . . . .width of a band in Hertz.

BASE STATION . . . . .transceiver in fixed location.

BEAM . . . . .directional type of CB antenna.

CARRIER . . . . .radio wave broadcast by a transmitter to carry the signal.

CHANNEL . . . . .specific frequency within a band. CB has 80 channels, in two bands.

CLARIFIER . . . . .control used on sideband radios for fine tuning to make speech intelligible.

COAX . . . . .coaxial cable. Shielded cable used to connect the transceiver to the antenna.

CRYSTAL . . . . .piece of quartz which vibrates at a constant frequency.

DC .direct current. The kind of current from a battery.

DECIBEL (dB) . . . . .unit which measures power ratio. There is a 3 dB difference between two signals one of which is twice as powerful as another.

DIODE . . . . .a rectifier that allows electricity to flow in one direction only.

DIPOLE . . . . .simple antenna.

DX .long-distance radio contact, usually caused by weather or atmospheric conditions.

ELECTROMAGNETIC WAVE . . . .any transmitted radio signal.

ERP . . . . .effective radiated power.

FADING . . . . .signal fluctuation as a result of atmospheric conditions.

FIELD STRENGTH METER . . . . . .instrument which measures the strength of a radio signal.

FM .frequency modulation.

FORWARD GAIN . . . . . . .the increase in signal strength that a beam-type antenna will give.

FREQUENCY . . . . .number of alterations per seconds of a signal. CB carriers alternate at 27 million per second, or 27 Megahertz (MHz).

GAIN . . . . .another word for amplification. If an amplifier increases a signal from 1 volt to 5 volts it has a gain of 5.

GROUND PLANE . . . . . . .omnidirectional antenna for base station use.

HERTZ . . . . .unit of frequency. Named after Heinrich Hertz.

INTEGRATED CIRCUIT . . .a single silicon chip containing a large number of transistors.

KILOHERTZ . . . . .1000 cycles of alternation per second. Written as kHz

LINEAR AMPLIFIER . . . . .used to boost output of a transceiver.

LOADING COIL . . . . .wire wound coil to increase electrical length of an antenna.

LSB . . . . .lower sideband

MEGA . . . . .prefix meaning millions as in Megahertz (MHz).

MICROVOLT . . . . .one-millionth of a volt.

MOBILE . . . . .radio installation in a vehicle.

MODULATE . . . . .add voice strength to carrier wave.

OHM . . . . .unit of electrical resistance.

OMNIDIRECTIONAL . . . .sending of a signal in all directions.

OSCILLATOR . . . . .generator of waves.

OUTPUT POWER . . . . . . .power generated by a transceiver measured in watts.

OVERMODULATING . . . .talking too loud, causing modulation to exceed 100% and resulting in distorted transmissions.

PTT . . . . .push to talk bar on microphone which allows transmission when pressed and reception when released.

PLL . . . . .circuit that generates the frequency and channels.

RADIO CHECK . . . . .short transmission to determine quality of signal.

RESISTOR . . . . .a device which impedes the flow of current.

RF . . . . .radio frequency.

ROTATOR . . . . .directs a beam antenna for maximum gain in one direction.

SELECTIVITY . . . . .ability of a receiver to ignore interference and transmissions not exactly on the tuned frequency.

SENSITIVITY . . . . .ability of a receiver to pick up weak signals.

SKIP . . . . .reflection of radio waves by the ionosphere, enabling DXing.

S-METER . . . . .meter on a CB to show strength of received transmissions.

SSB . . . . .single sideband.

SWR . . . . .Standing Wave Ratio. Comparison of forward and reflected power in an antenna system.

TRANSCEIVER . . . . .a radio set which can both transmit and receive.

TRANSFORMER . . . . .device to convert the voltage of an alternating current to a lower or higher value.

TVI . . . . .television interference caused by a radio transceiver.

UHF . . . . .ultra high frequency.

USB . . . . .upper sideband.

VOLT . . . . .unit of potential difference.

WATT . . . . .unit of electrical power.

WHIP . . . . .antenna.

Chapter 21

THE CB LANGUAGE

When CB first started in the U.K. the lingo was used a lot more than it is now, partly because in those days it was used to confuse the enemy. The pioneers of CB radio were operating illegally and so the slang was very useful.

There are mixed feelings amongst breakers about the lingo, but however anyone feels the lingo is here to stay even if there are breakers who disapprove and, after all, it is fun which is what CB is all about. I have condensed it quite a lot because most of the language drifted across from America and really isn't used in this country.

a

ADVERTISING . . . . . .police car with blue light flashing
AFFIRMATIVE . . . . .yes
ANKLE BITER . . . . .young child
ATTIC . . . . .channel 40

b

BACK DOOR . . . . .last CBer in convoy
BACK IN A SHORT SHORT . . . . .back in a minute
BACK OFF THE HAMMER . . . . .slow down
BACKSIDE . . . . .return trip
BAREFOOT . . . . .using rig without amplifier
BARLEY POP . . . . .beer
BASEMENT . . . . .channel 1
BASE 20 . . . . .CBer's home
BEAR . . . . .policeman
BEAR CAGE . . . . .police station
BEAR TRAP . . . . .radar trap
BEAVER . . . . .female
BENDING THE WINDOW . . . . .strong signal
BETWEEN THE SHEETS . . . . .sleeping
BIG 8's . . . . .best wishes
BIG 4 . . . . .yes
BIG SLAB . . . . .motorway
BIG WHEELS . . . . .truck
BIRD CAGE . . . . .airport
BLACKTOP . . . . .major road
BLEEDING . . . . .transmission affecting other channels
BOTTLE SHOP . . . . .pub
BLOWING MY EARS OFF . . . . .very strong signal
BOTTOM OF THE SHOP . . . . .channel 1
BOOTS ON . . . . .using linear amplifier
BREAK . . . . .request to speak
BREAKER ON THE SIDE . . .third party wants to join in
BREAKING CHANNEL . . . . .channel 19 / channel 14
BREAKING UP . . . . .signal becoming distorted
BROWN BOTTLE . . . . .beer

b

BUCKET-MOUTH . . . . .foul-mouthed breaker
BURNER . . . . .linear amplifier

c

CANDLES . . . . .years
CATCH YOU ON THE FLIP . . . . .see you on return trip /see you around

CB . . . . .Citizens Band
CHECK THE SEAT COVERS . . . . .look at women
CLEAN AND GREEN . . . . .road clear of police
C'MON . . . . .reply
COAX . . . . .co-axial cable
COME BACK . . . . .reply
CONVOY . . .a few CB equipped vehicles in procession
COPY . . . . .receive
COPY COPY . . . . .do you receive me
COPYING THE MAIL . . . . . .listening to other CB users
CRUMB SNATCHERS . . . . .children

d

DOUBLE L . . . . .landline/telephone
DOUGHNUT . . . . .roundabout
DOWN ON THE SIDE . . . .finished talking but listening
DO YOU COPY . . . . .do you understand
D.T.I. . . . . .Department of Trade and Industry
DX . . . . .long distance

e

EARS . . . . .antenna
EARS ON . . . . .on channel
EARWIG . . . . .listen to a conversation
EIGHTEEN WHEELER . . . . .articulated lorry.
EYEBALL . . . . .meet another CBer face to face
ELEVEN METRES . . . . .CB band

f

FAIRY LIGHTS . . . . .traffic lights
FINAL . . . . .last transmission
FLAT SIDE . . . . .horizontal polarisation, also-going to sleep

FLIP FLOP . . . . .return trip
FOR SURE . . . . .I agree totally
FOUR WHEELER . . . . .car or van
FRONT DOOR . . . . .lead vehicle in convoy

g

GARBAGE . . . . .interference
GET HORIZONTAL . . . . .go to sleep
GETTING OUT . . . . .broadcasting, being heard
GO BREAKER . . . . .permission to use channel
GOING DOWN . . . . .turning off the CB
GOLDEN NUMBERS . . . . .best wishes
GONE . . . . .ending transmission
GOOD BUDDY . . . . .fellow CBer
GOOD NUMBERS . . . . .best regards
GOT YOUR EARS ON . . . . .are you on channel
GREASY SPOON . . . . .cafe
GREEN SHIELD STAMPS . . . . .money

h

HAM . . . . .radio amateur
HAMMER DOWN . . . . .travelling fast
HANDLE . . . . .CB code name
HARVEY WALLBANGER . . . . .reckless driver
HASH . . . . .interference
HIGH NUMBERS . . . . .same as good numbers
HIT THE HAY . . . . .go to sleep
HOME TWENTY . . . . .base location
HORIZONTAL . . . . .asleep

i

IN A SHORT . . . . .in a minute
IN A SHORT SHORT . . . . .very soon
IS THAT A FOUR 10-4

j

JAM SANDWICH . . . . .police vehicle,
white with red stripe
JUICE . . . . .fuel
JUNK YARD . . . . .place of employment

k

KEYING THE MIC . . . . .pressing the mic key
KNOCK IT DOWN . . . . . . .go down to another channel
KNOCK IT UP . . . . .go up to another channel
KOJAK WITH A KODAK . . . . .police with radar

l

LADY BREAKER . . . . .female CBer
LANDLINE . . . . .telephone
LAY AN EYEBALL ON . . . . .meet in person
LINEAR . . . . .linear amplifier

m

MAGIC NUMBERS . . . . .best wishes
MAMA BEAR . . . . .policewoman
MAYDAY . . . . .distress signal
MIC . . . . .microphone
MIC FRIGHT . . . . .fear of transmitting
MOBILE . . . . .CB in vehicle
MODULATING . . . . .talking on the CB
MONITOR . . . . .listen in to a CB channel
MOTION LOTION . . . . .petrol
MUD . . . . .coffee
MUPPET CHANNELS . . . . .legal FM 40 channel CB
MUNCHIES . . . . .food
MUSHY . . . . .poor signal

n

NEGATIVE . . . . .no
NEGATIVE CONTACT . . . . .CBer does not respond
NEGATIVE COPY . . . . .no answer/not understood
NEGATORY . . . . .no
NINE . . . . .repeat last message/not understood

o

ON CHANNEL . . . . .on the air
ON THE SIDE . . . . .listening in
ON YOUR MUD FLAPS . . . . .following close behind
OUT . . . . .end transmission
OVER . . . . .transmission ended, your turn
OVER MODULATING . . . .signal too loud and distorted
OVER YOUR SHOULDER . . . . .behind you

p

PASS THE NUMBERS . . . . .best wishes
PEDAL TO THE METAL . . . . . . .accelerator to the floor
PERSONAL . . . . .your name
PICK A NUMBER . . . . .choose a channel
PLAIN WRAPPER . . . . .unmarked police car
PORCUPINE . . . . .vehicle with lots of antenna's
POSITIVE . . . . .yes
POUNDS . . . . .notches on the "S" meter
PULLING THE PLUG . . . . .switching off
PUSHING BIG WHEELS . . . . .driving a lorry
PUT YOUR BOOTS ON . . . . . . .turn on linear amplifier

Q

QUICKIE . . . . .short conversation
QUICK TRIP ROUND THE HORN . . .scanning the channels
QSL CARD . . . . .individual card to confirm long-distance copy

r

RADIO CHECK . . .report quality and strength of signal
RATCHET JAW . . . . .CBer who talks all the time
READ . . . . .hear
READING THE MAIL . . . . .listening to other CBers on channel

RIDING SHOTGUN . . .travelling in front passenger seat
RIG . . . . .CB radio
RINGING YOUR BELL . . . . .calling you
ROGER . . . . .yes affirmative
ROGER D . . . . .yes
ROGER BLEEP . . . . .a CB signal tone device
ROGER DODGE . . . . .yes
ROLLER SKATE . . . . .a small car
RUBBERBANDER . . . . . .green CB operator, new on CB
RUG RAT . . . . .young child
RUSSIAN WOODPECKER . . . .Russian long-range radar, it sounds like a woodpecker

s

SAY AGAIN . . . . .repeat transmission
SCAN THE CHANNELS . . . . . .looking for free channel

s

SEAT COVER .female passenger
SEE YOU IN A SHORT . . . . .see you in a minute
SHAKING THE WINDOWS . . . . .loud clear signal
SHORT SHORT . . . . .soon
SHOTGUN . . . . .male passenger
SKIP . . .reflections giving long-distance transmissions
SLAB . . . . .road
SMOKEY . . . . .police
SMOKEY BEAR . . . . .police
SMOKEY REPORT . . . . .police location
SMOKEY WITH A CAMERA . . . . .police using radar
SOCKS . . . . .small linear amplifier
SPAGHETTI . . . . .skip transmissions from Italy
SLASHED ON . . . . .interfered with by another CBer
SPLATTER . . . . . .interference from adjacent channels
SQUARE WHEELED . . . . .parked
SQUAWK BOX . . . . .CB set
STATION ON THE SIDE . . . . .CBer wanting to join conversation
STEPPED ON . . . . .transmission interrupted
STICK . . . . .mobile antenna
STROLLER . . . . .CBer using handheld
SUPER SLAB . . . . .motorway

t

TAKE IT DOWN/UP .move down /up to a chosen channel
TAKING PICTURES . . . . .police using radar
TAR . . . . .coffee
TEN TEN TILL WE DO IT AGAIN . . . . .sign off
TEN ROGER .message received
TIN CAN . . . . .CB set
TOILET MOUTH . . . . .obscene CBer
TOP OF THE SHOP . . . . .channel 40
TRASH . . . . .noise on channel
TURKEY . . . . .nuisance
TWENTY . . . . .location

t

TWIG . . . . .CB antenna
TWEAK . . . . .tune up
TWENTY . . . . .location
TWO WHEELER . . . . .motorcycle/bicycle
TWINS . . . . .duel antennas

v

VERTICAL . . . . .vertical ground plane antenna
VERTICAL SIDE . . . . .vertical polarisation

w

WALKED ALL OVER . . .blotted out by a more powerful signal
WALLPAPER . . . . .QSL cards
WALL TO WALL TREETOP TALL . . .good strong signal
WALL TO WALL SPAGHETTI . . . . . . .overcrowded with Italian skip
WALLY . . . . .idiot
WATERING HOLE . . . . .pub
WEARING BOOTS . . . . .using linear amplifier
WHAT AM I HITTING YOU WITH . . . . . . .how strong is my signal
WHAT ARE YOU PUSHING . . . . . . .what is your power?
WHAT'S YOUR 20 . . . . .what is your location
WHIP . . . . .mobile antenna
WIND UP . . . . .make fun of
WIPED OUT . . . . .overpowered by stronger signal
WORKS 20 . . . . .place of employment
WRAPPER . . . . .vehicle

x

XYL . . . . .wife

y

YL . . . . .young lady
YM . . . . .boyfriend
YOU CHOOSE IT I'LL USE IT . . . . .pick a channel

z

ZEDS . . . . .sleep
ZOO . . . . .police station

Chapter 22

THE CB LICENCE

Right, let's start by answering some of your questions about the licence for a CB radio.

DO I NEED A CB LICENCE

Unless you are using a CB radio under the supervision of another CB licence- holder, you will need to take out a CB licence in your own name. Please note that children under the age of 14 years cannot hold a CB licence.

HOW MUCH DOES A CB LICENCE COST

The licence cost 15 pounds per year no matter how many sets you use or how many of the three services you use.

HOW DO I GET A CB LICENCE

As from 1st April 1992, licence administration for Citizens Band Radio (and Amateur Radio) will be handled by the Radio Licensing Centre, Subscription Services Limited (SSL) Bristol. You can get a CB Radio Licence application form by simply writing to or telephoning the Radio Licensing Centre or the Radiocommunications Agency. Return the completed application form and correct fee to the Radio Licensing Centre. Within 14 days of the Radio Licensing Centre receiving your form and fee you will be sent your licence: (a Validation Document and booklet of licence conditions). The Radio Licensing Centre will accept crossed postal orders and cheques drawn on Bank/Building Societies, standing orders and direct debit. Cash payment will no longer be accepted.

HOW LONG DOES A LICENCE RUN FOR

All new CB licences will expire at the end of the month prior to the month of issue (e.g. a licence bought on 11 February 1990 will expire on 31 January 1991). This keeps the system simple and thus the licence fee down. So in the first year a CB licence will run between 11 and 12 months and then on renewal for 12 months.

DO I HAVE TO RE-APPLY EACH YEAR

No, the Radio Licensing Centre will send a computer produced renewal notice one month BEFORE payment is due. If you lose your renewal or do not receive one, re-apply on the initial application form, and tick the box marked "renewal".

IS ANYONE ELSE COVERED BY MY LICENCE

Yes, anyone may use your CB equipment but only under your "general supervision". Please remember that you are responsible for all use under your licence. General supervision will obviously apply to cases where the licensee is directly supervising a non-licensed person. Other areas of supervision are not so clear cut and examples of such cases are listed next.

1. EMPLOYEES

Equipment must be operated on, or in the immediate vicinity of, the licensee's premises, for example on a farm. Any vehicles leaving the licensee's premises would not be covered by his/her general supervision and the driver would have to be licensed separately. The CB licence covers the licensee to install and use CB. It does not cover the installation of equipment to be used by anyone. Minicab firms must ensure that ALL drivers are individually licensed. Should a driver leave, a new driver MUST be licensed separately. Taxi firms who operate CB radio have no special rights and must obey all conditions of the CB licence. In particular, they are not allowed to block channels and should not advertise their service on-air.

2. NON-EMPLOYEES

This covers any person who the licensee has control and authority over. The licensee must be willing to take full legal responsibility for any person supervised. For example, members of the same family may be covered by one licence even if operating away from home. Scouts or similar organisations need also only apply for one licence.

3. OLD PEOPLES NETWORKS

This applies to official OAP networks with a superintendent (or equivalent) who uses CB radio to contact the OAP's in the network at certain times of the day to check they are all right. This does not apply in cases where a CB club has provided a free CB set to an OAP for use in emergency or for general communication purposes.

FACT-FILE

Any correspondence on licensing should be addressed to the **Radio Licensing Centre, Subscription Services Limited, PO Box 885, Bristol, BS99 5LG.**

A telephone "Help Desk" **(0272) 258333** will be available 7 days a week between 8.30am and 10.00pm.

DO I GET A REFUND IF I STOP USING CB

No, refunds are not offered for CB licences. The Licence is not offered on a "time basis" or to cover specific items of equipment. The fee covers the cost of issue plus a contribution to the costs of the Radiocommunications Agency.

DO I NEED A LICENCE FOR "WALKIE TALKIES"

Yes, if they operate within the CB bands. Even low power hand-held CB radios must be covered by a licence. Note: Operation of walkie talkies at 49 MHz is licence exempt, however, the walkie talkies must conform to the performance specification MPT 1336.

DO I NEED A LICENCE FOR EACH SET OF CB FREQUENCIES

No, the CB licence allows you to operate on any (or all 3) of the sets of CB frequencies.

IS THE LICENCE REDUCED FOR ELDERLY OR DISABLED PEOPLE

No, the CB licence fee is the same for everyone. If concessionary fees were to be offered, it would increase the cost of running CB and so the licence fee for everyone.

Chapter 23

CB TROUBLESHOOTING

	1	2	3	4	5	6	7	8	9	10	11	12	13	14	15	16	17
No sound or Channel Light			●	●		●				●	●			●			●
No Channel Selector	●	●								●					●		
No or dim Channel Light														●	●		
Blowing fuses											●						●
No transmit		●					●	●	●	●	●			●			
Output meter reads lower than normal								●				●		●			●
Reports of weak signal strength			●					●			●	●		●			●
Reports of over modulating																●	
Channel Light on - no sound		●			●		●	●		●	●		●	●			
No voice reception					●		●			●			●				
Weak Receive nearby								●					●	●			

SOLUTIONS

1 Disengage channel 9/19

2 Check CB/PA switch

3 Check power lead connections

4 Check fuse

5 Check squelch adjustment

6 Check on/off switch

7 Change to active channel

8 Check antenna connections and cable

9 Fully depress push to talk (PTT) button

10 Check microphone connection/plug/lead

11 Check metal-to-metal ground connection

12 Check SWR

13 Check RF gain control or switch

14 Check battery power

15 Check LED switch or dim switch

16 Too close to microphone or speaking too loud

17 Check vehicle electrical system

Chapter 24

THE CITIZENS BAND LICENCE

These terms, Provisions and Limitations, shall be read as an integral part of the Citizens' Band Radio Licence

ESTABLISHMENT AND USE

2) The Licensee may
(a) establish any station or stations for Wireless Telegraphy formed only of telegraphy apparatus type approved or certified in accordance with either Department of Trade and Industry Performance Specifications MPT1320 or MPT1321 or MPT1333, or European Telecommunications Standards Institute Performance Specifications ETS 300 135 and marked on the front panel of such apparatus in accordance with the appropriate specification (a CB Station)

(b) use a CB Station for sending and receiving only:-

(i) messages in plain speech; and

(ii) signals of less than 2 1/2 seconds in length which are intended to call up any other CB Station;

(iii) signals of less than 1 second in length which are intended to identify a CB Station or to mark the end of a message by a CB Station.

LIMITATIONS ON USE

APPARATUS OF CB STATION

2) The wireless telegraphy apparatus forming a CB Station shall not be used:-

(a) unless it is operated on the frequencies specified in the appropriate Performance Specification

(b) if at any time, such use:-

(i) is likely to obstruct the Police, the Department of Trade and Industry's Radio Investigation Service or any other Government Body in the carrying out of their duties,

(ii) causes harmful interference to any other wireless telegraphy.

ANCILLARY EQUIPMENT

3)

(1) Except as provided at (2) below, only the following ancillary equipment may be connected between the antenna socket of the transceiver of a CB Station and it's antenna:-

(a) devices for suppressing interference;

(b) low pass filters;

(c) mechanically operated co-axial change over switches;

(d) where the station is formed of 934 MHz apparatus, antenna pre-amplifiers.

(2) A meter to measure the transmitting output power or Voltage Standing Wave Ratio (VSWR) of the antenna may be TEMPORARILY inserted between the antenna socket of the transceiver and the antenna for the sole purpose of establishing a CB Station or testing equipment provided that such meters are removed before a CB Station IS USED.

4)

(1) If a CB Station is formed of 26-27 MHz apparatus with provision for connection to an external antenna such apparatus shall only be connected to a single vertical omnidirectional monopole antenna, the driven element of which does not exceed 6.85 m in length or 55 mm in diameter, including any loading coils and associated circuitry and castings, but excluding any plates or radial wires or rods designed solely to act as a ground plane.

(2) If a CB Station is formed of 934 MHz apparatus with provision for connection to an external antenna such apparatus shall be connected only to an antenna consisting of not more than 12 elements, each of which does not exceed 17 cm in length.

(3) In all cases, if a CB Station is situated within 1 km of the boundary of an aerodrome the height of the antenna and any supporting mast or structure must not exceed 15 m.

OPERATORS OF STATION(S)

5) A CB Station shall only be used by:-
(a) the Licensee; or

(b) any person where such person's use is under the general supervision of the Licensee.

6) A CB Station:-

(a) shall not be established or used in any aircraft vehicle and shall not be established or used on board ship except with the consent of the Ship's Master.

(b) shall not be used for the purpose of advertising goods or services of any kind.

CONDITIONS OF THE LICENCE

7)
(1) Subject to the payment of a fee in the manner set out in sub-clause **(2)** below, the Licence shall continue in force from year to year until revoked by the Secretary of State.

(2) The Licensee shall pay to the Secretary of State before the anniversary of the Date of Issue in each year, the fee on renewal prescribed by the Regulations for the time being in force under sub-section 2(1) of the Act and on the payment of the fee the Secretary of State will issue to the Licensee a document in the form of the title page of this Licence (the "Validation Document") which will indicate the next date of renewal.

(3) If the Licensee does not pay the fee in the manner described in sub-clause 7(2), then the Licence shall expire at the end of the day before the relevant anniversary date of the Date of Issue.

(4) The Licensee shall surrender the Validation Document to the Secretary of State forthwith upon the revocation of the Licence.

(5) Where the Secretary of State exercises his power to revoke or vary all Citizen's Band Radio Licences, a general notice issued in accordance with Section 1 (4) of the Act to all holders of such Licences shall be published in the following manner:-
(a) by means of an authorised broadcast; or

(b) by means of an insertion in the London, Edinburgh or Belfast Gazettes.

8) The CB Station and this Licence shall be available for inspection on demand at any reasonable time by a any person authorised by the Secretary of State.

9) A CB Station shall be closed down at any time on the demand of any person authorised by the Secretary of State.

INTERPRETATION

11) In this Licence, except where stated otherwise, words and expressions shall have the same meaning as they have in the Act.
Issued on behalf of the Secretary of State for Trade and Industry

NOTES TO THE CB LICENCE

1) It is the responsibility of the Licensee to ensure that no unauthorised use of a CB Station takes place.

2) The CB Station may be used for business and other commercial activities subject to the limitations in sub-clause 5 (b) of this Licence.

3) Only equipment which conforms in all respects to the appropriate Department of Trade and Industry or European Telecommunications Standards Institute Performance Specification may be used. This Licence does not permit the use of equipment which has been modified in any way. In particular Licensees should note that this Licence does not allow use of equipment which has been modified:-

(a) to allow transmission or reception on other or additional frequencies;

or

(b) to produce a greater output power.
(However, terminals or other connecting points are permitted for the use of a voice synthesizer device to provide an aural indication of the selected channel. The voice synthesizer device shall not modulate the transmitter.)

4) Please note that this Licence does not permit:-

(a) use of the following equipment:

(i) Power Amplifiers (also known as “linear amplifiers”, “burners”, “boots” or “assistance”),

(ii) Antennas other than as specified in Clause 4 of this Licence,

(iii) Devices designed to encrypt to distort speech,

(iv) Devices to allow a CB Station to be interconnected with the Public Service Telephone Network.

(v) Devices to allow a CB Station to be remotely controlled,

(vi) Repeater stations.

(b) the following uses of licensed equipment;

(i) The retransmission of any pre-recorded material,

(ii) The retransmission of any broadcast material,

(iii) The transmission of music in any form.

5) Only mechanical antenna switches which do not use radio frequency energy in any way may be used.

6) Please note that the installation and use of a CB Station may be subject to local regulations and bye-laws. In particular the installation of antennas must conform to local planning restrictions. You may also need to obtain consent before using equipment covered by this Licence on private or public property.

7) When operating a CB Station, every effort shall be made to avoid causing interference to other electronic or electrical apparatus.

8) The Distribution of licences on behalf of the Secretary of State is undertaken by the Radio Licensing Centre, Subscription Services Ltd, PO Box 885, Bristol, BS99 5LG.

Application for a Citizens' Band Radio Licence

RA
Radiocommunications Agency

Please refer to the section notes overleaf and complete the following details in dark ink using BLOCK CAPITALS.

1 Details of Applicant

Title (Mr, Mrs, Miss, etc.) ______ Initials of forenames ______

Surname ______

Mailing Address ______

Post Code ______ Date of Birth __ __ 19__

Telephone (include STD) home ______ work ______

2 Future Renewal Payment

I wish to pay my renewal fee by:
Tick **one** only

1 Direct Debit ☐
2 Cheque/postal order ☐
3 Standing Order ☐

If you select **1** or **3** a bank payment instruction will be sent with the Validation Documents.

3 Declaration

I certify that the information I have given is correct to the best of my knowledge

Signed ______ Date ______

If you are under 18 years of age the signature of your parent or guardian is required.

Signed ______ Date ______

Name (BLOCK LETTERS) ______

If you provide false information you may invalidate your application and be liable to prosecution.

Information given in this form is subject to the provisions of the Data Protection Act, 1984 and will be used in accordance with the appropriate entry attributed to the Radiocommunications Agency in the register of Data Users and Computer Bureaux.

RADIO INVESTIGATION DISTRICT OFFICES

DISTRICT ADDRESS	COUNTY or AREA
RIS PO Box 2500 Belfast BT5 6RJ	Northern Ireland
RIS PO Box 2500 Bristol BS99 7LJ	Avon, Cornwall, Devon, Somerset
RIS PO Box 2500 Bury St. Edmunds IP33 2XB	Cambridgeshire, Norfolk, Suffolk.
RIS PO Box 2500 Canterbury CT1 1SY	Kent, East Sussex ,West Sussex.
RIS Glamorgan. PO Box 2500 Cardiff CF4 4YW	Dyfed, Gwent ,Powrys, Gloucestershire,
RIS PO Box 2500 Edinburgh EH11 4AQ	Borders Region, Fife Region, Grampion Region, Highland Region, Lothian Region, Tayside Region, Orkney and Shetney Islands.
RIS PO Box 2500 Glasgow G41 1QY	Dumfries and Galloway Region, Central Region ,Strathclyde Region.
RIS PO Box 2500 Guildford GU1 2BD	Berkshire, Surrey, Oxfordshire.
RIS PO Box 2500 Leeds LS1 5HX	Humberside, North Yorkshire, West Yorkshire.
RIS PO Box 2500 Northampton NN3 1TU	Bedfordshire, Leicestershire, Northamptonshire, Buckinghamshire.

DISTRICT ADDRESS	COUNTY or AREA
RIS **PO Box 2500** **Nottingham NG8 3SS**	Derbyshire Lincolnshire Nottinghamshire South Yorkshire
RIS **PO Box 2500** **Preston PR2 2YH**	Cumbria Greater Manchester Lancashire
RIS **PO Box 2500** **Southampton SO9 7TT**	Hampshire Wiltshire Dorset Isle of Wight Channels Islands
RIS **PO Box 2500** **Newcastle Under Lyme ST5 1TP**	Shropshire Staffordshire West Midlands
RIS **PO Box 2500** **Billingham TS23 2YT**	Cleveland Co. Durham Northumberland Tyne and Wear
RIS **PO Box 2500** **Warrington WA5 3AF**	Clwyd, Cheshire, Isle of Man, Merseyside, Gwynedd.
RIS **PO Box 2500** **Birmingham B3 1BZ**	Hereford and Worcester, Warwickshire, West Midlands.

LONDON DISTRICTS

DISTRICT ADDRESS	COUNTY AND AREA
NW LONDON	
RIS PO Box 2500 Stanmore Middlesex HA7 1BP	North West London Hertfordshire (see Annex 2 for relevant postcodes)
NE LONDON	
RIS PO Box 2500 Colchester CO2 7YZ	North East London Essex (see Annex 2 for relevant postcodes)
S LONDON	
RIS PO Box 2500 Whyteleaf Surrey CR3 0YL	London Boroughs South of the Thames

ANNEX 2

NW LONDON DISTRICT

N1,N4, N5, N8, N9, N13, N15, N16, N17, N18, N21, N22, E1, E2, E3, E4, E5, E6, E7, E8, E9, E10, E11, E12, E13, E14, E15, E16, E17, E18. EC1, EC2, EC3, EC4, WC1, WC2, EN1, EN2, EN3, EN7, EN8, EN9, CM5, CM14, CM16, RM1, RM2, RM3, RM4, RM5, RM6, RM7, RM8, RM9, RM10, RM11, RM12, RM13, RM14, RM15, RM16, RM17, RM18, IG1, IG2, IG3, IG4, IG5, IG6, IG7, IG9, IG10, IG11, LM13

NE LONDON DISTRICT

N2,N3,N6,N7,N10,N11, N12, N14, N19, N20, NW1, NW2, NW3, NW4, NW5, NW6, NW7, NW8, NW9, NW10, NW11, EN4, EN5, EN6, WI, W2, W3, W4, W5, W6, W7, W8, W9, W10, W11, W14, WD1,WD2, WD3, WD4, WD5, WD6, WD7, SW1, SW3, SW5, SW6, SW7, SW10, HA0, HA1, HA2, HA3, HA4, HA5, HA6, HA7, HA8, HA9, AL2, UB5, UB6, UB8, UB9, UB10.

CB LOG BOOK

DATE	TIME	STATION (HANDLE)	LOCATION 20	SIGNAL R.S.T.	CHANNEL	NOTES

CB LOG BOOK

DATE	TIME	STATION (HANDLE)	LOCATION 20	SIGNAL R.S.T.	CHANNEL	NOTES

CB LOG BOOK

DATE	TIME	STATION (HANDLE)	LOCATION 20	SIGNAL R.S.T.	CHANNEL	NOTES

CB LOG BOOK

DATE	TIME	STATION (HANDLE)	LOCATION 20	SIGNAL R.S.T.	CHANNEL	NOTES

CB LOG BOOK

DATE	TIME	STATION (HANDLE)	LOCATION 20	SIGNAL R.S.T.	CHANNEL	NOTES

CB LOG BOOK

DATE	TIME	STATION (HANDLE)	LOCATION 20	SIGNAL R.S.T.	CHANNEL	NOTES

CB LOG BOOK

DATE	TIME	STATION (HANDLE)	LOCATION 20	SIGNAL R.S.T.	CHANNEL	NOTES

CB LOG BOOK

DATE	TIME	STATION (HANDLE)	LOCATION 20	SIGNAL R.S.T.	CHANNEL	NOTES

CB LOG BOOK

DATE	TIME	STATION (HANDLE)	LOCATION 20	SIGNAL R.S.T.	CHANNEL	NOTES

CB LOG BOOK

DATE	TIME	STATION (HANDLE)	LOCATION 20	SIGNAL R.S.T.	CHANNEL	NOTES

CB LOG BOOK

DATE	TIME	STATION (HANDLE)	LOCATION 20	SIGNAL R.S.T.	CHANNEL	NOTES